Soil Information Systems

July '76

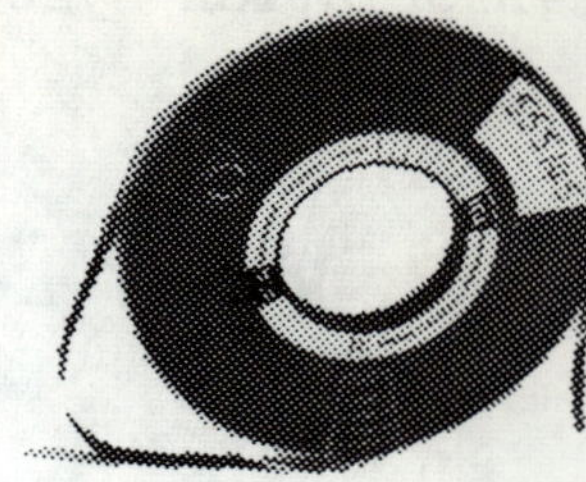

Soil Information Systems

Proceedings of the meeting of the ISSS Working Group on Soil Information Systems

Wageningen, the Netherlands
September 1-4, 1975

Editor: Stein W. Bie

pudoc

Centre for Agricultural Publishing and Documentation
Wageningen, the Netherlands
1975

ISBN 90-220-0592-5

Cover design: Pudoc, Wageningen

Printed in the Netherlands

Contents

Foreword

The First Meeting on Soil Information Systems in Wageningen, the Netherlands, on September 1-4, 1975 was organized by the Working Group on Soil Information Systems, under the work program of Commission V of the International Society of Soil Science. When this Working Group was being formed at the tenth Congress in Moscow in 1974, we were fully aware of its significance for the further development of soil science. We elected J. Schelling as convener of the Group. This resulted, within a very short period of time, in the first meeting reported here. With representatives of some fifteen countries from all continents meeting, the gathering contributed to the definitive organizational development of the Working Group. A comprehensive and up-to-date program was constructed, giving a solid and lasting basis for future work on soil information. Further meetings will follow: the regional Sub-Group for Australia in Canberra in 1976 and the whole group will meet again in Bulgaria in 1977. We have thus made provisions for the largest possible number of members of our Society to participate in laying the foundations of systematically organized work on the development of Soil Information Systems within the framework of the International Society of Soil Science.

It is natural for every research worker to believe his work is important; otherwise he would lack the motivation needed to submit to the stress and strain of research. However, in emphasizing the exceptional importance of work on the development of soil information systems, we have powerful arguments at our disposal. It is not an exaggeration to say that by developing the information systems in the field of soil science we are going to reach a radical turning point in our research methodology and our general approach to the problems of soil science.

By entering it into the system, the data on soils which we have been collecting and analysing hitherto in order to deal with a single, narrowly defined problem, will now acquire a permanent and universal value. At present, we can hardly foresee all the areas of application of such data. However, already the present degree of development of information systems, and the high cost of collecting and analysing soil samples, fully justify the view that in future every analysed profile, irrespective of the purpose for which it was recorded, should always be characterized by all the indices included in the system. The adaptation of methodology to the requirements of the information system will result in a definitive break with the old descriptive, largely subjective, methods of our field research. The logic of the information system cannot tolerate any vague or undefined notions. The new methodology has been followed by introduction of new words into soil science vocabulary. Good old soil profile descriptions have become INPUT; soil field maps have become manuscripts for DIGITIZING; soil description sheets became CODING FORMS; the field books stacks of PUNCH CARDS; new assistants appear carrying MAGTAPES instead soil augers and exchanging secret FORTRAN statements.

The new opportunities for research into the interdependence of phenomena and processes in soils, as well as studies of the dependence of soil properties on environmental factors on the basis of many data retrieved from the system, will largely alter the significance of experiment as the basic method in dealing with those problems. The use of information systems in the development of soil classification leads to a considerable rise in the accuracy of assignment to classes. It creates a basis for soil mapping and acts as an instrument for transfer of scientific knowledge and practical experience to ecologically similar conditions. Our information systems will become particularly valuable when they begin to include data on the yields of agricultural crops and forest plants. Such systems will become a precious source of information for agriculture and forestry, leading up to evaluations of world resources of food and raw materials.

Experience gained hitherto in the development of automated soil mapping with information systems has already sufficiently justified this research trend; such systems are already functioning successfully in some countries as a service to the farmer and agricultural industries. It would be very useful for this kind of activity to be connected and coordinated with the program of the Working Group on Remote Sensing, which also operates under Commission V.

With increasing pressure on land, and a wakening political and popular awareness of environmental qualities, data on soil resources have become valuable tools in planning. But planning decisions require a far quicker pace than that traditional to soil science. The shortening of response time in providing suitable data for land-use decisions can greatly improve the decisions themselves and give soil scientists a role of growing importance. We are not only scientists and thus responsible to our Alma Mater, but also the scientific guardians of soil resources and thus with significant social responsibility.

Although the activities leading to the development of information systems on soils are already reaching beyond the experimental phase, it may still be said that in this field we are at the start of the work to be done. This is why our Working Group was founded at the right time, since in this phase it is capable of exercising a powerful influence in the development of research. Undoubtedly, organization, initiative and concepts of work which this Working Group is about to develop will be largely responsible for the direction and rate of development in this scientific discipline. Hence this Working Group is faced with a great opportunity and great responsibility.

An important task of the Working Group is to organize education for a very wide group of soil scientists working in the fields of soil genesis, mapping, classification and geography, so as to reach at least the information level that will permit them to understand the significance and opportunities of the soils information system. Thus specialists working in this field can extend the circle of co-workers capable of assisting in all phases of the input, as well as of exploting data to the full. It seems to me that a manual in this field, written in an accessible manner, would help considerably in the pursuit of this aim, while in the near future we may think of establishing an international school, like the one on aerial photography in soil mapping we have in Enschede, the Netherlands.

The meeting signifies the beginning of organized actions of the International Society of Soil Science aimed at developing efficient soils information systems in all countries, a first step towards a world system.

The development of our International Society of Soil Science shows a clear trend of progressing from the initial phases, when the main form of activity was exchange of information, toward collaboration in research and resources. We are trying to deal with the problems of soil science by joint efforts contributing to the present struggle of mankind to increase the supplies of food and raw materials. In these noble endeavours, our Society has been successful for a long period of time in overcoming the limitations imposed by national borders. There is no reason to doubt its ability of acting as a nucleus for a worldwide system of information on soils. The Food and Agriculture Organization of the United Nations has devoted full attention to this activity. Within this organization, where our Secretary General, R. Dudal, holds an important position, this Working Group can most certainly find support, as well as a partner for an enterprise of this kind, just as Commission V, together with FAO, has recently completed its work on the soil map of the world.

It was a great pleasure for us to have gathered with such a significant task in mind in a country which has a great tradition and which has played an important role in developing soil science. Today it is one of the leading countries in the field of information on soils. We are very grateful indeed to our kind hosts, our Dutch colleagues, and to the convener of the Working Group, J. Schelling in particular, for their warm reception and excellent organization of this meeting.

M. Ćirić
Chairman Commission V
International Society of Soil Science
Šumarski fakultet
Zagrebačka 20
Sarajevo, Yugoslavia

Soil information systems in 1975

S.W. Bie and J. Schelling, Netherlands Soil Survey Institute, P.O. Box 98, Wageningen, The Netherlands

Introduction

The demand for specific, precise and rapid soil information is growing in our modern society. In our attempts to make optimal use of land, soil information plays an important role. Soil scientists have in recent years looked to electronic data processing as a tool that may allow more efficient handling of their soil data.

Operational computer systems have been with us for over two decades. In 1975, we are beginning to see the outline of information systems based on the computer technology now available. Soil scientists are not the first group to consider such systems. Geologists, geographers and cartographers have gained considerable experience with computer-based systems in related fields. But soil is a particularly complex and variable thing, and our progress has been a little slower. Also, soil scientists have taken it upon themselves to build soil information systems and have thus entered a field where they were not particularly well qualified.

In spite of all difficulties a considerable body of expertise on soil information systems now exists, scattered between many countries, poorly documented but rapidly growing. Here is a review.

Philosophies of soil information systems

The First Meeting on Soil Information Systems in September 1975 illustrated the multifaceted approach to computer-aided data handling. Many participants were scientists who had made their names in other branches of soil science: soil genesis, classification, micropedology, plant nutrition, soil terminology, soil mapping and land classification. Previously separated by subject barriers they found themselves with the computer in common. So computer-aided soil information systems must serve many purposes. In 1975, such systems are distinctly fragmentary.

Some have seen soil information systems as the most efficient filing system yet developed. Computer-based systems can extend or replace manual files. This is particularly important when files are to be merged, linking profile descriptions to analytical data, to cartographic files, results of field trials and to external non-soil files (climate, planning). Some comparisons and calculations on large data sets cannot be done without computer assistance. The arguments frequently used to support this are staffing or economic considerations: insufficient staff or the prohibitive costs of manual data handling, especially where manpower is scarce and expensive.

Also a number of scientific arguments are emerging. There are new ideas on sampling techniques for soil survey, soil classification and soil cartography, frequently statistically based. These scientists find an ally in the computer, which is a tool they were using anyway. They believe that the new data handling capabilities of the computer can lead to smaller information loss than with traditional methods.

So we believe that there are two philosophies at work. *Mechanization* attempts to translate current working routines into computer actions. The computer is used to store and retrieve more efficiently than is possible by manual methods. *Alternative automation* implements a new soil survey methodology, not yet accepted, with emphasis on statistical sampling design, numerical taxonomy and interpolation. This methodology is, almost incidentally, also very well suited to computer handling. In addition to storage and retrieval, alternative automation transfers many decisions in data processing to the computer through algorithms.

Whatever their origin and their aims, soil information systems have a number of modules in common. We have chosen to identify them as

- input
- data base management
- automated cartography and geographic information systems
- methodology and applications

No soil information system is complete without input, data base management and application modules; most will also have a cartography module.

Input

In the context of soil information systems, input modules fulfil three roles: to standardize the terminology used so that records collected by different observers at different times may be compared; to make input physically acceptable for computer handling, and to make different types of input compatible. The French-speaking world has made a major attempt to offer a standardized soil glossary. Arising from the work of the French interdisciplinary working group 'Informatique et Biosphere', it has been extended to include other French-speaking countries: Belgium, Canada and via ORSTOM (the French technical aid agency) also other parts of the Francophone world.

It is expected that FAO will release its own glossary in 1976. It remains to be seen whether such international glossaries will gain wide acceptance. Established national survey organizations frequently see little point in standardizing beyond national frontiers. International cooperation, for instance on food production, has not yet come far enough to promote active moves toward international standardization. Automation may change this.

Standardization at national level has proved successful. France, Canada, the United States and Sweden all have national standards operating, including for soil information systems.

To make traditional soil profile descriptions computer-acceptable can be a major task. Two philosophies have been developed. One is to accept the traditional descriptions, in free language, and to use a computer preprocessor to format the information automatically. The other is to ensure that the information is in a computer-acceptable form initially, by coding or recoding.

The first approach has been used by organizations where it is desirable to include many old records in the computer data base, and where manual encoding of old material is not feasible. One input module used by Netherlands Soil Survey Institute (Wageningen) accepts completely free language descriptions. It uses dictionaries to organize the information. The systems developed by ORSTOM (Bondy) and SOGREAH (Grenoble) in France similarly handle free language if attribute separators are manually included. Experience gained with such variable-format preprocessors suggests that information loss is low. Computer costs are higher than with coded input. Against this must be seen the possible advantage of allowing the pedologist to use hos own words to describe profiles and sites.

Other groups regard the discipline imposed by more rigid and frequently coded input forms as the best safeguard for consistent, high-quality information. An extreme form is the punch card of the portable card punch used for forest soil surveys in Sweden (College of Forestry, Uppsala). The system has proved most successful there. Others, mostly dealing with many more attributes, have found them less satisfactory. Many make use of various coding forms, which in different forms relate to the conventional 80-column punch cards. The most compact is probably that of INRA/ENSA in Montpellier (France) which has also been modified for use in Canada by the Soil Research Institute (Ottawa). The observer is expected to master a considerable number of codes, but many are self-explanatory. The Montpellier/Ottawa form allows very many attributes to be recorded on one form. The form in use by the Soil Survey of England and Wales (Harpenden) reflects the more summary profile descriptions made during routine surveys.

Not only is the input format important. Many organizations have attempted to reduce the transcription necessary to make the records computer-acceptable. The portable punch of the Uppsala system solves this directly. Others have tried optical input. The USDA Soil Conservation Service (Washington, D.C.) has long experience with the 'fill in'-optically read form. So far the capabilities of optical character readers (OCR) have been limited to numerals and a handful of letters. The USDA Soil Conservation Service's Montana group (Bozeman, Montana) has had good experience with the 'scratch out' multiple choice OCR form for field work. Similar forms have been successfully used for transcription of analytical sheets in the Netherlands (Wageningen) but was there considered unsuitable for field work. Many OCR forms are usually required for one profile, making them bulky for field use. OCR forms are usually processed by service bureaux with access to the expensive equipment required. An in-house mini-computer system now being implemented by the USDA Soil Conservation Service may make such forms more attractive.

The potential of OCR forms of various designs seems considerable. Whether they, or other forms of input, work in practice seems as much as part of the psychology used for their introduction in a survey as inherent in the design. There is thus no universally accepted input system available. What is ideal for one purpose or one organization is less suitable for another. In the rush to make data ready for computer handling many unsuitable forms have been designed. Field trials are important. Many surveyors have a particular emotional relation to their observations. Computer methods must ensure that this contact is not lost.

Data base management systems

Soil information systems have so far been largely experimental. Frequently designed by soil scientists, they have been intended for small-scale applications. Large production systems required by, for instance, national soil surveys, with facilities to handle a million soil profile descriptions or more and many maps, have yet to be developed.

There is good reason for choice of a simple file structure used by many small systems, namely that for small data sets sequential search is perfectly acceptable and probably most economic. The quantity of data is small, editing requirements are minimal and retrieval limited. All computer centres have file management systems available for such applications, and some of these have been used for soil information systems.

It is emerging, not the least through the experience of our geological colleagues, that large data sets require a different approach. The problem is difficult and relates to the quality, quantity and expected uses of the data, as well as to the user, his or her competence at electronic data processing and the specialized support staff available. Gordon (p.43) made a valuable contribution in illuminating these problems, and providing a framework for the analysis of the requirements a data base management system must meet in various circumstances.

Two types of data-base management systems seem particularly applicable to soil information systems. One is required for small systems, with limited data sets (often only one file), another for large, multifile systems.

Through a series of examples, van den Driessche (p.17) illustrated that soil data can take different structures even when limited to well-defined variables. In larger soil information systems, we are often faced with files on different aspects of the soil and its environment, with different data structures.

Soil scientists have a number of operational small systems. Again SOGREAH (Grenoble) and ORSTOM (Bondy) in France have taken a pragmatic approach to this problem. On a somewhat larger scale, the Canadian Soil Research Institute (Ottawa) has embarked on a number of computer files on soil names and descriptions, cartographic information, management and soil degradation. The USDA Soil Conservation Service also builds files on soil series and interpretation. Reports from Sweden (Uppsala), Bulgaria (N. Poushkarov Institute, Sofia) and Rumania (Research Institute for Soil Science, Bucharest) likewise indicate files being built of analytical, morphological and interpretation data, again with simple file structure and file management.

For soil scientists, there is no data-base management system available off the shelf. This is a problem of 1975; we do not believe it will exist in 1980. New machine-independent data-base management systems with a friendly user face are being developed. GRASP (US Geological Survey, Reston, Va) is one of the first of this generation. Such systems will shortly be implemented for soil information and may bring the computer much closer to the small and apprehensive computer user.

For large systems, soil scientists must recognize that this is a field for specialist information scientists. The amount of expertise required exceeds that normally available within a soil survey organization. Commercial systems exist, many carry a price tag exceeding US $100 000. They also require considerable computer expertise to operate.

Jeffery (p.36) has outlined the inherent problems of multifile systems and the philosophies applied to solve them. Within a very few years, some large soil survey organizations will be forced to chose a data-base management system, for no functional soil information system can operate without efficient management. For many, it will be the computing centre that will decide, and it is unlikely that we shall all end up with the same system. The work of our geological colleagues may well prove relevant, for data-base management has been their main topic of conversation over the last few years. G-EXEC (Institute of Geological Sciences, London, GB) is probably the most advanced system developed by geologists. We may see DASCH (Geological Survey of Lower Saxony, Hanover) used for soil information systems in West Germany. Others will chose generalized data-base management systems, such as SYSTEM 2000 or TOTAL.

What seems essential for us soil scientists is to establish an exchange format, irrespective of the data-base management system we finally acquire. Again we may want to look to the geologists, who are now (late 1975) experimenting with the FILEMATCH format developed by University of Western Ontario (London, Ont., Canada) and the Institute of Geological Sciences (London, GB). FILEMATCH will provide a link between all major geological data-base management systems currently used in Europe and North America.

Automated cartography and geographical information systems

Maps must be drawn. They provide geographical information, for instance on soil and thus constitute a particular kind of data file.

Computer-aided cartography began just over ten years ago, and at two levels. On the one hand an English group (that was later to become the Experimental Cartography Unit, London) demonstrated that one can translate lines into digital coordinates and reproduce this file on a plotter, achieving acceptable cartographic results. About the same time, the work on discrete cell maps began, of which the Graphics Laboratory of Harvard University (Cambridge, Mass., US) became an exponent, through its lineprinter mapping system SYMAP. SYMAP also had facilities for creating maps from point data without boundaries being defined.

The application of automated cartography to soil mapping has been slow. Little real progress has been made with two exceptions. The first is the introduction of interactive editing facilities offered by storage cathode ray tubes linked to the computer that also controls the digitizing table and the plotter. The second is that philosophies of data-base management have penetrated to cartographic files. File organization has improved dramatically in recent years. It has been inspired by work in other fields, by the US Bureau of the Census (Washington, D.C.), Harvard University and others.

Until recently very few soil survey organizations had ventured into automated cartography. The Soil Survey of England and Wales (Harpenden) had conducted trials with the Experimental Cartography Unit (London), publishing a high-quality soil drainage map in 1973. Webster (of that Survey) and Massey (University of Melbourne, Australia) experimented with line-printer maps and linked aspects of numerical classification to soil cartography. This work has recently been extended by de Gruijter (Wageningen), also using colours. Another lineprinter system, MIADS, is used by the USDA Soil Conservation Service (Fort Worth, Texas) for the production of multicoloured derived (interpretive) soil maps.

The slow progress is associated with three problems:
- the inability to manipulate with polygons
- the high investment required for high-quality systems
- the doubtful economics of production

The solution of the first problem came from the electronics industry. Early work on soil problems by Cook (CSIRO Division of Land Use Research, Canberra, Australia) had not been put into production, and the work by the Experimental Cartography Unit (London), also on soil maps for FAO, had proved feasible but time-consuming. However, a self-made system by the Soil Research Institute in Ottawa (Canada) seems promising. A number of firms operating in the electronics industry turned their attention to cartography. Systems, such as that now being delivered to USDA Soil Conservation Service (Johnson, p.48) are spin-offs from investments made for a much larger market than thematic cartography. The investment needed remains high, but the economics of production now look more favourable with increased machine speeds.

Van Zuijlen (p.52) outlined the basic technical principles of computer-aided cartography. From this, and the experience in soil survey organizations, we suggest that the following types of systems are of particular interest for soil cartography:
- for quick low-cost maps: a line-printer mapping system implemented on a large computer and able to make contour maps, proximal (nearest neighbour) maps and conformant maps (grid version of existing maps), e.g. SYMAP (Harvard University). We foresee the lineprinter soon being replaced by the electrostatic plotter giving higher quality output.
- An automated cartography system for high-quality maps, with a minicomputer controlling a digitizer, an interactive CRT unit, a check plotter (e.g. drum plotter) and a high-quality flat-bed plotter (or the use of such a plotter). At least a small disc and a magnetic tape unit are necessary. The software must be capable of interactive editing and polygon manipulation.

The cartographic file has two functions: to draw maps (soil

maps or derived maps), and to act as a data base for geographical queries. Many have been content with mechanized map drawing, but in terms of an information system the latter aspect is equally important. FAO has explored this in cooperation with the Experimental Cartography Unit (London) on the soil map of South America.

The cartographic file as a reference file is an important development that has been greatly stimulated by the Canada Geographic Information System (Dept of Lands, Ottawa). Technical problems remain before a smaller soil survey organization can hope to use the methods developed there. But this use of digital cartographic data holds much promise.

On its own, automated cartography may to-day be difficult to justify economically (note that the period chosen for writing off equipment is critical to the estimates). Only repeated use of cartographic files can at present justify the high investment required. In future, aspects of geographical information handling may form the major usage of these files.

Methodology and applications

Relatively little has been done to link research methods developed for experimental purposes to practical soil information systems. There were two broad groups of interest represented at the Working Group Meeting, indicating that the designers of soil information systems have their roots firmly in research and in practical applications.

One group emphasized the role of information systems for more efficient collection and processing of soil information, improving the quality of statements about soil. It was possible the first time that colleagues who had previously worked in relatively narrow fields within soil science, extended their vision -- in public -- to the basic problem of soil survey methodology. The papers of Ćirić (p.10), Teaci (p.61), Schelling (p.13), Webster (p.65), and Bie (p.73) raised increasingly critical questions about many of the basic and traditional methods of collecting and processing soil data. They feel that many of the traditional methods have developed within the limitations of manual data handling, and that with the advent of computerized information systems many of these constraints are removed. Beckett (Oxford University, GB), whose work has long centred on this theme, concurred and expressed concern over conflicts that may arise if data collected according to some assumptions (e.g. free survey) be processed by soil information systems applying algorithms less perceptive than those of the surveyors.

Not all data are of the same quality (Webster, p.65), and poor data should not be included in the data base. We may want to include a reliability tag with a record.

Here are the beginnings of an important new discussion on soil survey methodology, precipitated partly by the new data-handling facilities of soil information systems. There are two main approaches: one, particularly stressed by our French colleagues (e.g. Ruellan, ENSA, Rennes), attempts to interpret soil-forming processes quantitatively, thereby increasing our understanding of the soil landscape and thus enhancing our powers of predicting soil conditions. The other approach regards the soil landscape as a source of uncertainty and attempts by statistics to resolve it in an optimal way. Exponents of this approach advocate numerical classification and spatial geostatistics (partly developed at Fontainebleau in France by mining engineers) for the most powerful predictions. An interesting variant is a mixed traditional/statistical approach to soil mapping presented by Ragg (Soil Survey of Scotland, Edinburgh). Within a statistical sampling framework the surveyor is free to add information in the form of subjective vectors. The vectors influence the map construction done in the computer. This is an imaginative attempt to marry human perception and computer power.

The other main trend in the application of soil information systems concerned soil and land suitability estimates and advisory work. Teaci (p.61) and Garbouchev & Sadovski (Sofia) outlined the ambitious plans of the Rumanians and Bulgarians to use their soil information systems for planning and advisory purposes. They wish to link management parameters to soil data. The Sofia group is developing a cybernetic model intended to act as a source of continuously optimal information on land use, even within a crop season. They foresee the interfacing of external computer files on meteorology and management to the soil information system.

Flach noted the attempts towards this end within the USDA Soil Conservation Service, for instance, with meteorological files, and stressed that the public interest in the environment is now such that only efficient soil information systems can provide the data needed rapidly enough for further evaluation.

Within the United Nations, FAO has to collate information from many corners of the world. FAO clearly sees its role as a catalyst (Riquier, p.79), trying to encourage soil and site information to be presented in a uniform way. Their own work has concentrated on using soil information to aid fertilizer use in developing countries. They have extended this to the farm advisory service aimed at the small peasant farmer. FAO will shortly propose an exchange format for this type of information. FAO clearly feels that the technical advantages of computer use in this kind of research in developing countries outweigh any social and economic objections that may be raised.

Further international cooperation to ensure compatibility between environmental information systems was proposed by Lange (UNESCO, Paris) and Tomlinson (International Geographical Union, Ottawa, Canada). IGU also has much data on information systems in use for other types of spatial analysis, and publishes an extensive catalogue on geographical information systems and related subjects.

Conclusion

No large soil information system is yet in practical use, so it is difficult to assess the role soil information systems will play in the collection and use of soil data. We think that the impact will be significant, both for soil survey methodology and for applications. The experience gained with the first generation of systems now being constructed, will greatly ease the problems for soil survey organizations that have not yet taken the plunge.

We will do well to look beyond soil science for inspiration. Our geological, geographical and cartographical colleagues face much the same problems and have often found solutions. The information scientists must also become more closely involved in our work. Soil information systems are too important to be left to soil scientists. They must interface with other information on our natural resources and on the land-use planning.

This book is a step towards rapid dissemination of current know-how on soil information systems. We sincerely hope that it will rapidly be outdated.

Acknowledgments

The Working Group on Soil Information Systems of the International Society of Soil Science extends its thanks to the publishers of this book for their spirited approach to the request for rapid publication of the ephemeral information on soil information systems. The Editor acknowledges with gratitude the help given by J.C. Rigg, J. Vermeulen and Mrs M. Elzenga of Pudoc towards this end.

The role of soil information systems

J. Schelling, Netherlands Soil Survey Institute, P.O.Box 98, Wageningen, The Netherlands

What is an information system?

An information system is a system consisting of procedures, machines, information carriers and data, that can be used to execute the necessary data and information processes.

The information system, the people and their environment

An information system is built to be used by a specific group of people, an organization, in a specific environment.

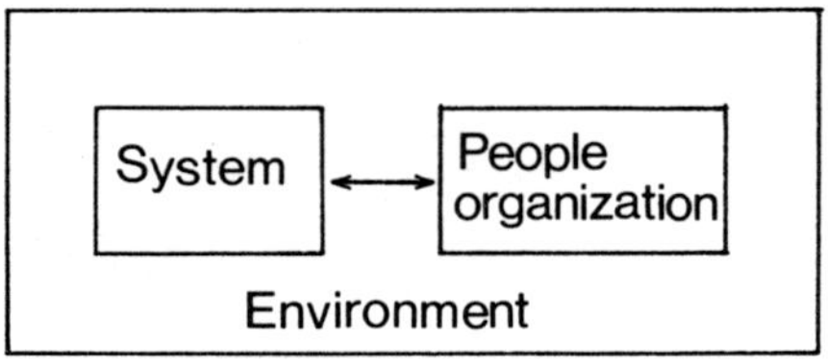

The role of the system for the organization is service. The organization must be willing to accept new methods, and this is one of the most difficult tasks in introduction. A step by step construction and introduction of an automated system is therefore to be preferred, people can gain experience with limited parts, integrate them in their present work, and possibly save time and labour and increase their scope.

The kind of organization under consideration can be characterized by its tasks, staff, budget, performance and prospects. These in turn are influenced by the environment of the organization.

A small group of experts, with lower echelon staff of limited training, with few facilities and cheap manpower in a developing country, will have quite different needs from a highly trained group of experts, where capital is abundant, and labour is expensive.

In this paper the soil information system will be considered as a 'black box', able to perform whatever we want.

The role of soil information systems can be considered under different aspects:

- economics and efficiency
- quality, form and usefulness of the products
- scientific aspects
- man-machine relations
- role of groups of information systems.

Economics and efficiency

Hard facts on the economics of information systems are scarce. Developing or buying systems is very expensive, e.g. the development, without full implementation, of the Dutch WIA system costs roughly one million guilders. We can expect that the total costs will increase. This will partly be balanced by increased capabilities, and speed and quality of the system. At this stage every system is an experiment, without any guarantee of success.

The efficiency can increase considerably:

- we can store and handle very

large data sets, that cannot be handled simultaneously in any other way

- these data can be available and accessible immediately
- once input and programs are operative, the system cannot produce any mistakes
- the system works fast.

The system is possibly economic in use when the same data have to undergo frequent handling, because the same data are used several times for different purposes, or because the same data are used to produce a lot of different products. This can be so when many maps are derived from the same data in large land consolidation or development schemes. If labour is expensive, the system may be interesting as a labour-saving device.

A quite different aspect of efficiency is that the design of the system requires a careful analysis of present and future procedures, products and their qualities. This in itself may have such a stimulatory effect, that it can become one of the major achievements, even if the automated system would prove economically unfeasible.

Quality and usefulness

We all make mistakes, and the input of our system usually contains many mistakes, inaccuracies and inconsistencies. In the traditional methods, many of these are corrected while the surveyor is using his own data. But with machine operation, this is impossible. Data validation is one of the cornerstones of a good system.

Another important influence of the design of the system is, that because it requires unambiguous information, the organization is forced to use uniform terminology and methods.

For interchange between different soil systems, this may prove to be one of the major difficulties. To develop a common format for interchange of information is not impossible, but I fear that the lack of an internationally accepted terminology and common methods, linking information systems on an international level, may be virtually impossible. This drawback might on the other hand act as a catalyst to terminological standardization.

The use of information systems will have a considerable influence on the *form* in which we produce our output. This has a lot to do with the efficiency aspects, too.

The results of any data selection, and calculation or interpretation can be produced within seconds on a cathode ray tube, on hardcopy, a quick plotter or line printer. This means that direct answers to specific questions are instantly available. We can probably give our customers better products for every single step in which some of our information is used. This may lead to a new type of throw-away maps, as sensational as the birth of the paperback for publishing.

Models, e.g. in hydrology and planning, need digital information as their 'food', and systems can produce this easily. Since many planning and construction tasks use models, this aspect is rapidly becoming important.

Because calculations and statistics come into easy reach, these will be stimulated, and this may have a considerable influence on our work. Quality control may become another possible aspect of the system. This is prerequisite for improvement of our methods, one of the most neglected areas of research in soil survey.

Scientific development

The design and construction of an information system confronts us with all aspects of soil survey and interpretation, and thus the principles of present methods are considered from new angles. We should have the courage to slaughter our holy cows, and have a close look at all stages and aspects of soil survey, to explore new avenues.

Data reduction, a must for cartography, is for an information system not such a pressing problem. This means that data compression, and the loss of information that it brings about, can be diminished considerably. Combined with the increased capabilities for data manipulation and calculation, we must expect considerable changes in our methods.

Man-machine relations

Let us consider some possible negative influences first. All the fears, dislikes and frustrations that cropped up during the industrial revolution, come to life again when we try to automate soil information systems.

Essential for the system is such boring routine work as punching and digitizing. It will be essential to try to automate as much as possible of these tasks.

Some people see the system as a dumb machine that frustrates man by its stupidity. The system can only be as sophisticated as the people that designed it. That means that only those parts of our knowledge that can be formulated explicitly, can be used in the design. All the subtile considerations that our mind can execute, are out of reach for the system. In some respects the system may be rather crude and primitive.

Because of the easy accessibility of the data, use is easy, but misuse is less evident. Therefore explicit control procedures are essential.

The legal aspects of data ownership, and responsibilities are still obscure, and require further study.

Some of the positive aspects of man-machine relations have been considered earlier. Data validation can improve input considerably. A lot of tedious and time consuming jobs can be taken over by the system, e.g. endless consultation of soil profile descriptions to make specific derived maps, with all chances of mistakes and misinterpretations.

The guiding principle in system development should be the following: the system should be a tool, serving the people, and all solutions should be geared to their ease of use, acceptance and well-being.

If, for instance, we can choose between an elaborate coding system, with thousands of codes that all users have to learn, and the feeding to the machine of free text, the system performing necessary transformations, we prefer the latter.

Conflicts of interest that concern job satisfaction may pose difficult problems.

Man-machine interaction should be easy, in order to exploit the system, certainly when people have their doubts about its merits. Therefore I expect that interactive use of a system, in which the machine produces quick answers to the questions of the users, and thus stimulates him to do further steps, based on this information, will promote use, and let people acquaint themselves with new capabilities. This means that easy access to the system, by simple user-oriented commands, will be important especially in the first stage of use. A simple data base management system, like GRASP (from US Geological Survey), that can be used after five minutes introduction, can help us to overcome the user's initial fears.

Role of groups of information systems

The international aspects of soil information systems will be discussed by others. I will not even try to open this discussion now.

Let us consider the role of a soil information system in a specific area, for instance a country, in relation to other information systems in the same area. The possible uses increase greatly when systems can be linked, because then the different data can all be combined easily.

In the Netherlands we need a combination of specific soil and groundwater data about cultivated land for planning in every municipality. Once we have all data in digital form, and we can put them into the same format, it is an easy job. This indicates that, once we have several systems available simultaneously and the exchange formats, and communication are taken care of, the value of the separate systems increases considerably. In the information jungle, where everybody uses different machines, languages and software, this coordination may be a difficult task.

Data input and data structure

R. van den Driessche, Office de la Recherche Scientifique et Technique Outre-Mer, 70 route d'Aulnay, Bondy, France

Introduction

Before signing a contract with a hardware supplier, a user must choose among manual input, optical reading and automated capture of the information intended for the computer. Most people use a keyboard to enter data on paper tape, punched cards, magnetic tape, or disks. Some users have their alphanumeric data typewritten and loaded by an optical reader. The same readers can be fed with handwritten digits and a few letters. In the field, there is perhaps an opportunity for a few specific environmental items to be received by a satellite from data collection platforms, at regular intervals. In the laboratory, Technicons or other analysers can record the 'results' in a variety of ways (tape, cards, etc.) for input.

But the question is not only HOW but also WHAT shall we put in. This brief presentation of the problems of data will take the form of 4 case studies.

Case Study 1: class texture

More often than not a textural class is recorded in the field. Most manuals recommend a specific triangle of reference; some glossaries postpone such a choice between triangles and let the surveyor make a choice. The latter increases cost when it comes to data processing, unless processing is a mere editing of profiles. Statistical treatment of the profiles has gained wide interest in the last decade. Thus the question is raised: Which of the five following manipulations should texture undergo?

In the field, the item appears as a nominal variable. Usually this is the lot of a synthetic item. No relation other than that of equivalence being applicable to let's say 11 or 25 names of an unordered list. One of the names is recorded in the field, as such, or as a symbol.

e.g. sandy clay loam scl 8 16

Statistical treatment of nominal variables can take the form of a contingency table analysis (Dixon, 1964), of a similarity index based on probabilities (Goodall, 1966), or others. My involvement with rank-order statistics has led me to put the item in twice:

.sandy clay loam.scl.

is punched anywhere in the horizon record. The system is then instructed by a dictionary to treat the first item as a sand class number and the second as a clay class number. Both are ordinal variables. Here the relation 'greater than' holds. Upon retrieval, tape output is in the I-type format:

9 6

A very simple modification of the dictionary would transform the duplicated item into two interval variables (or ratio variables) sand percentage and clay percentage:

60 27

These percentages, by the way,

are the coordinates from the center point of the SCL polygon on the sand and clay percentage scales in the FAO triangle; they can be used -- with other items -- namely to compute a multi-level dissimilarity index between profiles.

A similar modification would transform the 'sandy clay loam. scl' input into class limits on the sand and clay percentage scales:

4480 2035

these, in turn, are input for other programs by format card:

44-80% sand and 20-35% clay.

From a cost point of view, the most appealing input seems to be symbol

.scl. punched

anywhere and transformed by the program into percentage sand and clay.

Case Study 2: particle size analysis

Let us consider 8 classes used in routine work by Hubert for INEAC in the Mahagi in the 1950'ies: LT 2, 2-20, 20-50, 50-100, 100-250, 250-500, 500-1000, 1000-2000 μm. The percentages were delivered by the laboratory to one decimal place:

36.8 7.0 5.1 4.2 10.6 10.4 20.8 5.1%

Is this not unwanted accuracy? Would rounded-off percentages not suffice? If so, one could allocate 8 specified 2-column fields to the percentages or one could keypunch

.37C1.7C2.5C3.4C4.11C5.10C6
.21C7.5C8.

anywhere in the record of v variables.

Case Study 3: field pH

The next case study on field pH is presented as a block diagram with all the alternatives open to the planner of a soil information system. (see p.19)

Case Study 4: profile descriptions

We recently created a reference file for profile descriptions from 230 sites in central Africa (Sys 1972), using techniques developed by Van den Driessche *et al.* (1974a, b; 1975a, b). 1. The master cards were in fixed format: one card for the profile number, ten cards for the ten recorded depths (1cm, 5cm, 12cm, 25cm, 41cm, 61cm, 85cm, 113cm, 145cm, 181cm).

PROFIL/NO 92/
- HORIZON/HRZ 5/41CM//

2. The data cards were in free format: English names of the observed and analysed items were punched, continuously, disregarding any order within the record, with just a period as separator. No blank space nor code for missing data:

SANDY CLAY.SC.7,5YR.5/.6.
MEDIUM.BLOCKY STRUCTURE.
.MODERATE.FIRM.ROOTS.35F1.
3F2.5F3.10F4.21F5.12F6.14F7.
0F8.0RE.0,4C.0,07N.5,0PH
1CA.0,3K.6T.POINT 92.DEPTH 5.

The same horizon was punched twice (or duplicated) when coverage extended over two depths of the above list. When maximum depth was not attained in a profile, the master card remained so to retrieve a record of missing data codes. Conversely, for the deepest profiles, a horizon may have been overlooked.

Per profile: 23 field variables + 16 laboratory variables at 10 depths = 390 variables. Two or three cards were needed for each depth record, exactly 3530 cards, master cards included.

3. We ran the data with our program DGY (69K) on our UNIVAC 1108 (192K) and listed the profiles. A magnetic tape was loaded simultaneously with codes. This happened in flip-flop mode and

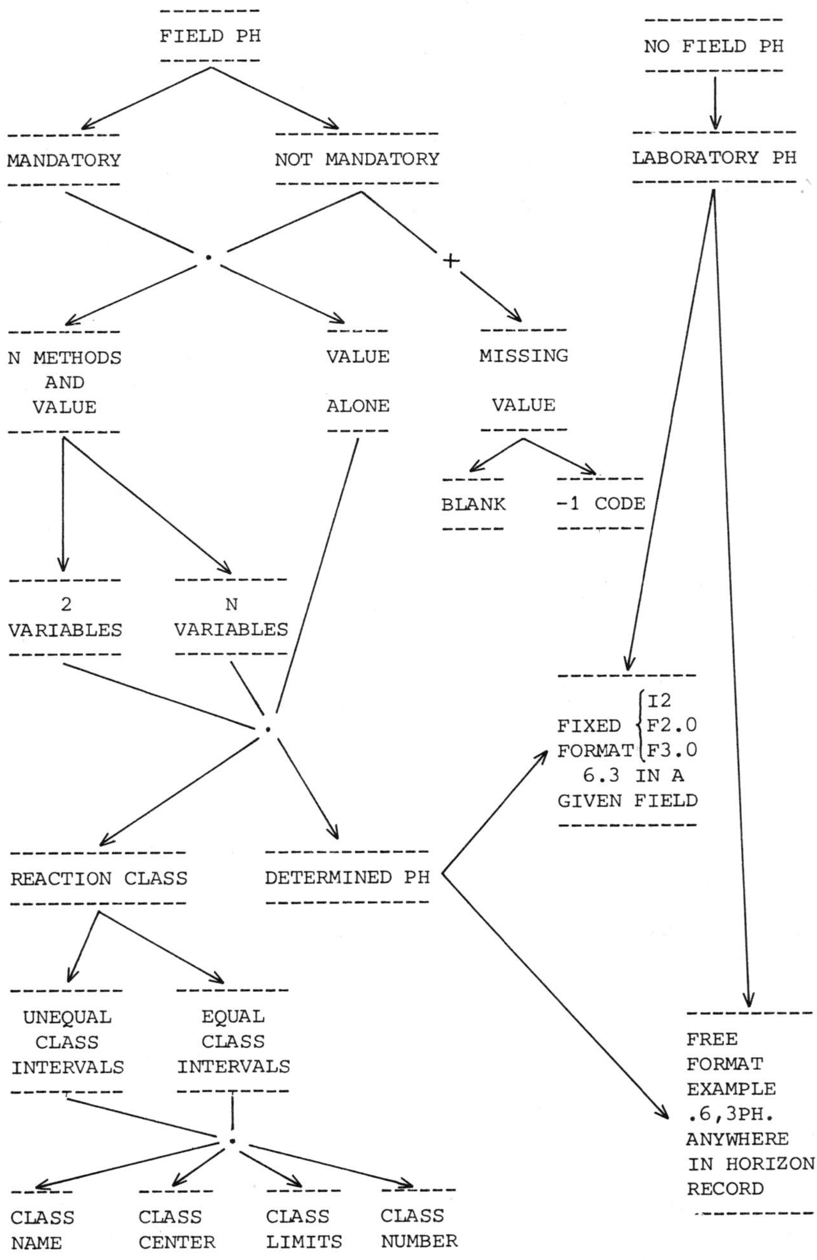
FIELD PH
NO FIELD PH
MANDATORY
NOT MANDATORY
LABORATORY PH
N METHODS AND VALUE
VALUE ALONE
MISSING VALUE
BLANK
-1 CODE
2 VARIABLES
N VARIABLES
FIXED FORMAT I2 F2.0 F3.0
6.3 IN A GIVEN FIELD
REACTION CLASS
DETERMINED PH
UNEQUAL CLASS INTERVALS
EQUAL CLASS INTERVALS
FREE FORMAT EXAMPLE .6,3PH. ANYWHERE IN HORIZON RECORD
CLASS NAME
CLASS CENTER
CLASS LIMITS
CLASS NUMBER

took 10 min (CPU) for a bundle of 100 profiles.

Unexpected, redundant, or unseparated items, or mis-spellings were annotated by the computer. The corresponding decks were then amended, and corrected profiles were run again (same time: 6 s/profile). Editing soil profiles took 1 s/profile. Example of an edited record:

```
- HORIZON/HRZ 5//
POINT 92.AT 41CM.SANDY CLAY.SC.
MODERATE.MEDIUM.BLOCKY
STRUCTURE.FIRM.
7,5YR.5/.6.
ROOTS.
35F1.3F2.5F3.10F4.21F5.12F6.
14F7.0F8.0RE.
0,4C.0,07N.5,0PH.1CA.0,3K.6T.
```

4. Our input procedure required previous loading of a dictionary, in which every item had an equivalent code number. Program RGY (68K) was used for that (28 s), but only once. A second, equivalent, dictionary afforded automated translation, simply by changing a control card.
5. Spatial and temporal referencing of the sites, plus environmental items (vegetation, climate, ..., center points of satellite imagery), plus soil classification data, were keypunched in the same way (24 variables) and loaded onto a second tape set. Large data files could be accommodated by the programs: 100 000 records of 500 variables.
6. Records were retrieved by input of a Boolean expression including less than 64 conditions written in plain language or in code (SGYT 80K and SGY 74K Fortran programs). Retrieval time is 1 s/profile.
7. For a geosearch input (Fortran program COOR 11K with worldwide coverage), 4 cards were needed inside an area rectangle which included the geographic latitude and longitude coordinates (accuracy $\pm$ 1'). The purpose was threefold: printout in English; data matrix on tape; spatial referencing of ground truth, satellite and Slar imagery was graphically displayed on the lineprinter (CARTO program 15K in Fortran), with a $24^{\circ}/18^{\circ}$ range and a 5'/3' accuracy for latitude and longitude, respectively.
8. All F-type formats were accepted by our Fortran statistical programs: one-level dissimilarity (DISSIM 47K), multi-level dissimilarity (+MERGE 8K), agglomerative clustering (AGGLOM 89K), multivariate identification of profiles (IDENT 12K), multivariate ordering of the sites, and the S' test (4K).

Conclusion

Through these 4 case studies I have attempted to show some alternatives in the input of punched items. The need for standardization across borders is felt. However, there is a danger in postponing the technical problems (hardware, software, statistics) until agreement is reached on an international list of priority items (point data + area data + ephemeral data + interpretive data) for exchange of data files. If we want to scrutinize various existing data base management systems, we must be prepared to reformat our data to achieve a standard test. The processing of foreign files, as in our central Africa reference file, is a stimulating experiment.

References

Dixon, W.J. ed. (1964). BMD Biomedical computer programs. 1st ed., UCLA, Los Angeles, 585 p.
Goodall, D. (1966). A new similarity index based on probabilities. Biometrics, 22, 882-907

Sys, C. (1972). Caractérisation morphologique et physico-chimique de profils types de l'Afrique centrale. INEAC, Brussels, 497 p.
Van den Driessche, R., Garcia Gomez, A., Giey, A., (1974a). Un système informatique en langage naturel: application didactique aux coloris ISCC-NBS. Init. doc. tech., ORSTOM, 25, Paris, 120 p.
Van den Driessche, R., Garcia Gomez, A., Giey, A. (1974b). SIDA Satellite imagery descriptors analysis. ORSTOM, Bondy, 165 p.
Van den Driessche, R., Garcia Gomez, A., Aubry, A.M. (1975a). A multi-level dissimilarity index between soil profiles. Cah. ORSTOM sér. pédol., 13, 2, in press
Van den Driessche, R., Garcia Gomez, A., Giey, A., Aubry, A.M. (1975b). POSEIDON Procédures opérationelles en statistique et informatique pour données en langage naturel. ORSTOM, Bondy, in press

Input of data to soil information systems

B. Kloosterman, Soil Research Institute, Canada Department of Agriculture, Central Experimental Farm, Ottawa, Ont., Canada

The operational success or failure of a soil information system can be largely dependent on data input. Inadequate procedures will reflect in a reluctance by field staff to collect data. This can result in missing data, error and incomplete computer runs.

The purpose of this paper is to review data entry operationally. The route of data movement will be traced from the source or point of observation to storage of data in a computer-compatible form with error reduced to an acceptable level. Error correction will be considered an aspect of data entry, although it is often associated with data manipulation.

The paper will deal with the various steps of data input, methods of data entry, methods of data collection, devices used for data conversion and considerations for data entry design. Each topic will be discussed from a pedological aspect. Comments will focus on field data because the collection of laboratory data is more straightforward. Other contributions will deal with cartographic data.

Steps of Data Input

Data Collection

This task can be accomplished by direct and indirect means. Direct observation or measurement includes things that can be seen or observed at the site. Indirect measurements can result from remote-sensing operations.

Field check or editing

The data collector at the earliest opportunity checks his data for consistency and completeness. He often adds in data related to the subject but not collected at the site. This may include climatic, geological or vegetative data available from other sources. An effort also is made to clarify any observations that do not adequately represent the object under investigation.

Transcription

The data have to be prepared or converted to a form 'readable' by the computer. Often field observations are transferred to specially prepared forms that can be key punched or optically read. This may involve free-text transfer or conversion of data to a series of characters or codes. Often this work is done by students or part-time clerical personnel. This can result in error resulting from misinterpretation, omission and transcription. Checking is necessary before the documents are key punched or read.

Key punching

The coded document is submitted for key punching onto card, tape or disk. This output is checked for accuracy by a verification procedure.

Computer submission

Large computers are normally used to edit, update and store information in files. Most file management systems do a certain amount of consistency checking automatically. Other data checks are usually designed specifically for the type of data. Although most errors can be corrected by clerical staff, some corrections have to be made by the originator of the data.

Methods of Data Entry

Data are generally modified for purpose of input, although some systems accept free format records. Often data collection procedures are dictated by the method of data entry selected. Computer costs can vary considerably with the method used.

Free-format entry

Data in free format are perhaps the least structured of all data types. The originator of the data makes few if any concessions to the computer. The most common form of free-format entry is one that utilizes a set vocabulary of key words which serves to organize data in the file. A special character is usually used to delimit variables on input. Data may be retained as character sets or converted to codes. Generally much computer time is needed and frequency of error can be excessive since the computer depends on pattern recognition for file organization.

Fixed-format entry

The most common mode of input is fixed-format entry. The data are assigned specific spaces on a card or record. Therefore, the computer depends on position orientation to organize the file. Special forms are necessary to record field data. Theoretically error should be less frequent than for free-format entry but poor form design and layout results in utilization problems and error. A well designed form is important for the field investigator.

Coded data entry

This method is treated separately since special considerations are involved that do not apply to fixed-format entry. Coded data may be entered in free or fixed-format modes. Once entered into the computer, processing can be very efficient on the raw coded data. However, if decoded output is desired, reference tables have to be available to the retrieval program with a resulting increase in retrieval costs. Input documents can be very compact, since coding often generalizes the data. This is an advantage to the person in the field, but is offset by the problem of converting information to codes at the data source. This is in itself a potential source of error if the data collector becomes too familiar with the codes and starts to rely on memory. Surveyors often resent having to spend time coding their data.

Multiple-choice entry

Procedures based on this method are characterized by ease of use, low frequency of error, legibility regardless of the data collector, and is moderately easy to convert to a computer-compatible form. However, it is mandatory that the form is well defined and comprehensive for collection of the data. Failing this, the method will greatly inhibit and restrict collection. Another drawback is that forms tend to be bulky and heavy for a field party.

As all methods of entry have some distinct advantages, it may be advantageous to incorporate more than one method of data entry. Often information that lends itself easily to coding may be set up for multiple choice equally well. But a free format section to cover comments and reporting unique observations is often desirable.

Data-recording methods

The most important data-recording methods that are or could be used for input in pedological surveys will be discussed. Again a defined data-recording procedure may incorporate one or more methods.

Field notebooks

Historically, this has been the most common method of recording data. Although it gives the data collector complete freedom in what he observes and records, the approach is not well suited to standardization of records. Although these documents can be transcribed to a format compatible with electronic data processing, completeness, legibility and organization of the data has a considerable bearing on accuracy.

Data-recording cards

With a gradual systemization of survey procedures came a need for more rigid recording techniques. Many organizations designed custom-made cards to facilitate the recording of data. These are created with the field pedologist in mind and are fairly easy to transcribe to a computer-compatible form. Legibility and completeness still remain potential problems in using these forms.

Punch cards

Essentially these consist of coding forms printed directly on computer cards. The cards are semi-perforated so that an entry is recorded by punching out the desired perforation with a pencil or other sharp instrument. Consequently, the data cards are immediately ready as input for a card reader. Use of these cards represents one of the shortest routes between data collection and the computer. The main disadvantages are that the amount of data that can be contained on a card is limited, mistakes generally mean rejection of the entire card and they require careful handling as bent cards are difficult to process in a card reader.

Coding forms

These are specially designed documents on which data are assigned to specific fields on the card. Data may be recorded in a free format, coded or quantitative mode. This type of form is often not well suited to either data collection or key punching. They seem best suited to fairly small data sets that require a minimum number of reference tables for codes.

Mark-sense forms

This method of data collection has been used only recently in the earth sciences. They are best suited for multiple-choice approach since responses are position-dependent on the form. Recent advances enable carefully written characters to be recognized as well. Although the documents tend to be bulky, they are immediately available for input to the computer. Since input is accomplished by scanning the document, special precautions have to be taken in the field to ensure that records remain clean. The popularity and practicality of this technique is increasing as sensing capability expands and costs decrease.

Multiple-choice forms

These are much the same as mark-sense forms in organization. The main difference is that these documents are key-punched for input to the computer. They are more suitable for use in the field since no special precautions are necessary in handling them.

Voice recording
Some field men have attempted to record their observations on a cassette tape which is later transcribed in the office to coding forms or some other medium. Apart from the bulkiness and care that a tape recorder requires, considerable self-discipline is required to record information in a logical sequence for easy transcription and editing. In addition, there is always the chance of discovering the tape is blank after description of several profiles.

Digital coding
Some of the forestry staff in the western United States have been using an electronic version of the multiple-choice form in the field. However, the instrument is more commonly found in the merchandizing industry. The device consists of a recorder with a sensor shaped like a pen that detects the pattern of a number of parallel lines of varying thickness. The pattern is converted to a numeric value or code. With a specially designed plastic template, one can get a large number of codes onto a sheet the size of a page: hence the possibility of a multiple-choice form. In addition, a small keyboard enables input of numeric information. Data are stored on a cassette and an interface will transfer it to computing facilities. With present capabilities, these instruments are restricted to small data sets.

Data from instruments
Increasingly, instruments in the field and in the laboratory are fitted with analogue-to-digital convertors for direct transfer of data from the instrument to a digital form. Although conversion equipment of this nature can be fairly expensive, it can be justified if sufficient data are generated. If the measuring instrument is calibrated properly and is functioning normally, this method of data collection is less prone to error than any other.

Devices used for data conversion

There are a variety of devices used to convert data to a computer compatible form.

Card keypunch
This old standby of data entry is still in common use and by no means outdated. For simple data sets, where formats are fairly constant, this still may be the most cost-effective way of converting data. If the cards are interpreted, they are easy to read manually, By indexing the data set, the cards can be manipulated mechanically before input to a central computer. Cards can also be easily manipulated, added to or deleted from. For large sets and volumes of data the old card-keypunch is hardly ideal.

Paper tape
The paper tape keypunch also has its place, although it faces stronger competition from alternatives than the card punch. Paper tape offers the distinct advantage of transmitting data conveniently over communications networks. However even though it is a cheap method of input, its limitations in editing, error correction and speed should be recognized.

Key to tape
I refer here to the direct transfer of information from the

keyboard to a 7 or 9 track industry-compatible magnetic tape. One of the advantages of this method is that data can be directly fed to the computer. Other advantages include greater flexibility, speed, variable lengths of record and greater freedom in formating. Such systems are well suited to large complex data sets and large volumes. This type of equipment can also be used as device for remote data entry with data communication links. The other variations of this basic concept are cassette and discette or floppy disk systems. The only disadvantage is that special readers are required to transfer data to the main computer.

Interactive terminal
In this instance, the data-entry station is on line either to a minicomputer or to a large time-sharing system. As data enter the computer directly, editing can be accomplished on input. The keypunch operator can be notified immediately of any mistake. Also data that have been entered can be displayed almost instantaneously as required. If this device could be stationed so that the operator is under the supervision of the data recorder, the data should be of high quality. This approach is powerful and attractive but the question remains whether soil data warrant this power and expense.

Optical reader
This group of data-entry systems has progressed far from the day of special forms and eletromagnetic pencils. This mode of entry is attractive since there is no human intervention between data collection and the computer. Although some character recognition is now possible, most data we still collected on special forms, which are read by a reader. With decreasing costs, this type of data conversion will play a larger role in future.

Considerations of data input

There are a number of things that one must bear in mind when confronted with the problem of data entry to a computer system.

Integrity of the data collector
Often little consideration is given to the person who is responsible for data recording. Historically the decision for what data were to be collected and how, were up to the person in the field or his supervisor. Data collection procedures could vary with the circumstances of the survey. With the advent of data-recording cards and forms, he often uses one form for a variety of purposes. Consequently data-recording can no longer be completely tailored to the task. In addition, the data recorder can become so attached to the form by inserting codes into blanks that his attention to the phenomenon he is observing is restricted to the elements on the form. Another problem is that coding forms can make the data so impersonal that the person becomes detached from it. It is imperative that the person sees his data come out of the system in readable form as soon as possible after submission. In data-input design the data collector must be considered to give him enough freedom to develop his knowledge and perception.

Minimize transcription
Every operation involved in passing data from the point of observation to the computer is a source of possible error. All effort should be made to minimize the number of steps. Records that

can be read directly into the computer have high appeal. Each organization, therefore, must weigh carefully the alternative routes and, keeping economics in mind, design the shortest pathway fot data.

Data accuracy

The method of data collection also has a large bearing on data accuracy. The cost of reducing error to an acceptable level can be high but may increase exponentially as error is reduced to zero. One must accept a certain percentage of error and try to live with it. The consequence on the computer is that since data are subject to more manipulation, the erroneous results can have a greater impact. A second point is that because of volume and the carrier i.e. data are stored on magnetic tape or disk, errors often go undetected for longer. Sometimes, data are not discovered to be erroneous until a certain type of retrieval is made and the program fails. The data may humanly not be in error but a unique feature of the file management program may take exception to the data.

Procedures for error detection and correction must be carefully adhered to ensure data accuracy. Also input procedures should be kept as straight forward as possible to minimize problems. Many problems cannot be anticipated and thus account for long periods of time before systems are fully operational.

Volume and cost

The attitude for several years has been, if it is 'data', put them into the computer and the data will be much more valuable and useful. There are perhaps some data that should not be put into the computer because the volume and cost, relative to usefulness, do not warrant it.

Assuming that the data should go into the computer, the question still remains of how. The broad criteria are volume and cost, cost being broken down into manpower, input, form preparation, etc. The problem becomes quite complex. A general rule, however, is the greater the volume, the more justifiable are costs for special forms and special input. The cost of a record decreases with increase in number. High costs per record that are traced must be carefully studied. For example, one may decide that for a given data-collection exercise where there is a large number of records but a small set of variables that special input procedures are not warranted, and that input can be in a free format. In fact, computer activity is high because the free-format data have to be deciphered before storing in a computer file. The computer activity could be decreased substantially by investment in special forms.

Another yeardstick that may be used is the greater the number of data items in a record the more desirable is a sophisticated input like mark sense input. Again the task and resources have to be carefully weighed up by each agency.

Interim access to data

The task of the data recorder often does not end with initial collection of the information. He will use the information to write a report, interpret and evaluate. The data must remain relatively accessible to him. For free-format input, it usually is sufficient simply to copy the documents for the scientist for his reference until information comes from the computer. As data input becomes more specialized and the data becomes an abstract array of numbers and symbols, results must be returned in an easily readable form to the originator as soon as possible. The weeks and months often required to get the information out is unacceptable. This dilemma favours input procedures that can be directly utilized by the computer. Interim access is a serious problem since, if not solved, recorders will often not cooperate or cooperate reluctantly with the originator.

Data standardization and quality control

If we are committed to adopting computer systems, data standardization is almost a must, especially if routine procedures are to apply to many people or agencies. In Canada, where our system applies to the whole country, it was necessary, even with a national classification system, to define all variables and categories used to make sure that values and codes would be applied uniformly. Forms and documents were also largely standardized. If not, systems would have to be modified for every new project. Without standardization, broad general studies and comparisons would be impossible because incompatibility of data. On the other hand, data standardization is restrictive and procedures have to be defined to handle unusual circumstances.

Important to the success of data standardization is quality control. Some organization has to be set up to monitor the total operation, and assure that rules and regulations are implemented uniformly. It also has to provide feed-back mechanism though which errors, problems, and unusual circumstances can be considered, acted upon and the result conveyed back to the people in the field. Communications in Canada is through a working group as well as visits and memos as required.

Stability of the system

Another point that should be covered is the length of time a system operates without modification. Disruptions in stability can occur, for instance by a change in classification system, or input procedures. This is possibly one of the most perplexing problems. In a large complex system, a small trivial change in input can have impact throughout the entire system.

For this reason, it is highly desirable, but not always possible, to freeze the system for a significant period of time. In the interim, evidence can be collected to warrant an overhaul of procedures. At the end of the freeze, changes can be adopted, retraining and reprogramming effort initiated. Often data that exist in the system have to be updated to remain compatible with incoming data after modification. Since this can be an expensive proposition, especially for a large system, it may be desirable to maintain dated systems. This means that the updated system processes only updated information. This may be a viable alternative since access to a dated data set decreases with time. However, if more than two or three systems have to be supported, it may be advisable to update past data.

Completeness of data set

The biggest single problem encountered in designing a data entry system is defining the constellation of variables to include in a data set. This is especially true if a number of survey organizations are involved. There are several factors that result in the collection of different variables. The orientation of the survey possibly has the largest bearing on the set of variable. A survey on an agricultural area would differ from a survey conducted on a national park. The details of a survey would also determine the number of variables. There is also a considerable difference in the data collected in different regions, because the rate of soil-forming processes varies considerably. In Canada, an attempt has been made to define an all-inclusive data set. This work has gone on for almost 4 years and has now become fairly stable with a total of about 750 variables. Having defined all variables, we have considerable flexibility in what information can be collected in any given survey. New or unique variables may be included under 'special notes'. If 'special notes' variables prove to be collected frequently enough, they can be incorporated with the standard set when revisions are made.

<u>What data to include</u>

There are two aspects to this question. The first deals with what variables should be included in a data set. A general rule is that any data that can be calculated or derived from the existing data set should not be included. A second rule states that data that are not unique to the data set should be considered at a higher level. Regional variables are good examples of this.

The second aspect is concerned with what data from a total survey should be entered into a soil information system. Should all routine data be stored or should omly a detailed representative be entered? In Canada, the soil-data file contains detailed representative data sets. The soil description file contains both information collected day by day and abstract profile descriptions. In the Ottawa survey, almost 100 000 observations were recorded. Each observation contained about 12 items. The purpose of this data is to assist the pedologist in defining the soil individuals. It has yet to be proved whether this information can all be utilized or whether the manipulation possible on the computer warrant the cost of putting these data into an information system.

<u>Mobilization of data input</u>

Developing a data-entry system from conception to implementation can take several years. This is the result of a number of factors. It takes a reasonable length of time and number of revisions before input documents are ready for a field trial. In Canada, a working group was formed to develop the forms. Field trials were preceded by a pilot project in which, among othet things, field forms were tested. After training sessions in the office and in the field, the forms were released for field trials for a fixed period of time.

One of the most important tasks was to condition field men to the concept of data collection for computer input. They must be convinced that the effort is worthwhile for them. The pilot project was useful in illustrating ways that the computer could be used. Although the coding forms were designed for use in the field, they often proved to be filled in by transcription in the office. After a two-year period, feedback from users indicated that the forms were unsuitable for field use. The documents were then completely revised to a multiple-choice format. These are currently being used in the field and evidence so far suggests that they are better suited to field work that the first-generation forms.

Conclusion

It should be emphasized that there is no ideal for data entry. There are many considerations that make the analysis of the problem difficult. However some broad rules can be drawn up.

1. Do not disrupt the data-gathering process.
2. Assist the scientist in broadening his knowledge and perception.
3. Keep the route between the field and computer as short as possible.
4. Minimize errors.
5. Provide continuous access to the data.
6. Tailor procedures to the budget of the agency.

Data entry is the front door to a data system. It is important that it is always open to allow information to pass freely for reference and analysis.

Large soil data sets and their manipulation

P. Jongen, SOGREAH Consulting Engineers, 47 Ave. Marie-Reynoard, Grenoble, France

Introduction

Soil science is a relatively recent natural science, which integrates many variables, some requiring knowledge of other closely related sciences. Its main objectives are the study of soil origins and characteristics, the explanation of the phenomena which occur in soils and the definition of interactions between soil and human or natural activities in the soil.

A soil scientist will have a vast amount of qualitative and quantitative data on hand, which will have to be sorted, interpreted and classified for fundamental or applied research purposes.

In soil surveying, it is difficult for the soil scientist to synthesise all the data collected and it is evident that by using data processing methods, calculations can be automatic, thereby avoiding long and tiring routine: calculation of mean, median, minimum and maximum, standard deviation, and correlations, and factorial analysis to mention a few.

By applying mathematical methods to this data, a very complex set of facts can be studied with comparative ease where the human mind alone would have great difficulty. Data are processed with greater objectivity.

General principles for collecting and processing information

In order to handle the ordering of a continuously increasing amount of information and to plan future systems that can be improved without going back on all that has been done in the past, a few fundamental principles have to be borne in mind and certain difficulties have to be accepted.

- The establishment of data must meet a certain number of criteria, defined by Girard (1975) as being collected, structured and standardized. Information obtained from an observation, description or sample must be recorded in a coherent and comprehensive notation system such that, for example, accurate geographical coordinates can be obtained. This data must be recorded on physical carriers compatible with data processing techniques (e.g. punched cards). In this way, the data can be checked, interrogated and statistically processed at a later date.

- Provision must be made for the 'language' difficulty. Although this problem does not arise to any great extent for quantitative data, owing to the fair degree of agreement on the more general methods of analysis, the same can not be said for the vocabulary used for descriptive variables. However, the establishment of technical glossaries has initiated a ceratin international harmonization of the various systems of nomenclature. Achievement so far suggests that this difficulty is not insurmountable and that it is often possible to identify a given concept, using terms formulated from different systems. A simple example illustrates the correlation between French and English terms dealing with soil moisture conditions: the French terms 'sec' and 'frais' are taken to have a single English equivalent 'dry' which is a single concept defining a given state in the soil.

- Prerequisite for data processing is that the problems must be defined according to a strict and simple logic for both fundamental and applied research. In the applied field, a program must be designed such that the maximum generalization is achieved, i.e. combination of criteria is taken into account in order to avoid having too limited a framework. This analytical stage requires deep thought and targets clearly defined. For fundamental research, a prerequisite is the definition of working hypotheses to be applied in mathematical methods. The purpose of this first analysis is to define problems raised by application of these methods to soil data. The model would then be determined after an objective and complete observation of the data table. As indicated by Ruellan (1975), these are methods of descriptive statistics, or 'data analysis'.

- If more detailed quantification is required, the problem of weighting variables has to be overcome, thereby creating the problem of defining the hierarchical nature of these variables. For example, if a variable (e.g. the thickness of a stony layer) has to be weighted as a function of a given depth in the profile, the weight attached to this variable amy vary according to whether reference is to a genetic classification (generally low weight) or to an irrigability classification (higher weight, since irrigability is related to water retention coefficient and permeability). Useful information could be drawn from the work by Riquier (1972).

- Finally, the systems to be proposed must make provision for operational factors, a point often ignored. It is essential to have the correct frame of mind to save effort and time at each stage of collection, interrogation, processing and generation of data.

Schematic organization of a processing network

The manipulation of large sets of data requires the organization of a processing network based on the following points:
- Experience shows that data collection and input onto an adequate support is an onerous and costly stage, which must be undertaken with the utmost care.
- The basic data set is processed by a suitable program such that it can be reconstituted in the form of a clear description of the profile and of analysis tables. Operational models exist that produce an ordered arrangement of the various descriptive and quantitative variables translated on request into various natural languages.
- The following stage selects the variables and classifies them in one or several homogeneous subsets with a view to constructing new, more elaborate, data sets for more specific processing:
* Such a program must be capable of selecting and grouping a succession of horizons within a set of profiles, or a set of genetic horizons comparable with several profiles, or sets of character-

istics identified by their depth in the profile in accoedance with various combinations of criteria.
* The interrogation method generally practised (van den Driessche, 1974; Sogreah, 1972) uses, by analogy with Boolean algebra, the questions AND and NOR and the functions OR or OR NOT. The questions AND and NOR are used for variable selection processing, whereas these questions in association with the functions OR give a branched graph, each node representing a variable and the edges the functions. This type of processing may be aimed, for example, at classifying the land in terms of irrigability.
* For analysis, it is advantageous to be able to validate data, either qualitative or quantitative, in order to obtain reliable information.
* In systems design, to avoid excessive congestion of the storage facilities, the program must be designed for establishing tables to permit sequencing of the variables and recording of the rank of variables in the recordings.
* Data retrieval results in the formation of a series of homogeneous subsets, which consist of either elaborated data sets for further processing or a soil classification for agricultural development studies as described above (by adopting, for example, US Bureau of Reclamation standards).
* The transfer of subsets can be achieved on printed lists, on punched cards, on magnetic tape or, if required, on magnetic discs. Several subsets may be transferred onto the same carrier by placing them end to end, but care should be taken to draw up an automatic labelling system to find the start of each subset on the carrier rapidly.
- Specific operations, such as statistical calculations and tables, frquency studies, crossed dispersions, calculation and plotting of data on a graph (e.g. texture triangle), correlations, factorial analyses, are made on the previously defined subsets. There is also the prospect of introducing data from a different discipline, agronomy for example.
In conclusion, it would seem essential to make provision for the use of these types of program in a general manner in order to attain targets fixed by the soil scientists: more thorough interpretation of the vast mass of raw data, to store the data and the results obtained, and to supply the user with easily interpretable results. Another point of view is that the use of statistical methods makes the soil scientist more precise in his own field observations and gives a more critical outlook on quantitative data derived from the laboratory. This processing network seems to satisfy the needs of the practical soil scientist and the research worker.

Data management

It may be asked whether this large volume of data will have to be preserved. If so, in what form? How could such data be used in the future? And in what ways could it be rendered easily accessible for later work? Of the present needs of users, the following can be quoted:
- in fundamental research, the identification and explanation of genetic phenomena by the compilation of much data on one or several soil characteristics;
- in the experiment, the generalization of results for regions of comparable soil character, under identical climatic conditions;
- in agricultural planning, the

definition of land development parameters from qualitative or quantitative soil data.

These questions lead to the study of the problem of data storage and management in order to meet the requirements of various users. The establishment of data banks will enable this information to be grouped together. On this subject, a study group organised by Girard (1975) made the following proposals:
- The information collected by a data bank may be one of three types:
* raw, without any calculation or interpretation
* elaborated, implying an analysis of processed raw data
* referenced, i.e. including the site in coordinates and a certain number of essential references on the origin, the type of information and the various operations performed.
- If one conceives a central organization collecting data from local banks and wishing to offer potential users the maximum of information, it is the last type of data, i.e. referenced data, which seems most attractive since it does not complicate the storage method and is easily and rapidly available to the user.
- Local data banks are linked to this central organization. These local banks operate independently, collecting raw data, exploiting and processing it with their own programs but they are obliged to supply the central organization with referenced data. For local centres, the first managerial aspect is the organization of data in the storage facilities in order to restrict sequential search to the same level as that used with a dictionary, where the first letter of a word is sought by the first syllable. The second managerial aspect concerns the method of interrogation, which must be an easily understandable language for users unpractised in data processing. This method of management requires a constant exchange of data. Such exchanges already exist in certain countries, although it would be better if action were more coordinated world-wide.

Conclusions

Action should be taken to obtain up-to-date information on the various curent research projects and a *catalogue* should be drawn up as proposed by the 'Working Group on Soil Information Systems'. Computer specialists know from experience that much time is needed to perfect processing systems and it would be a good thing to pool the experience acquired by various working groups. The development of data-processing methods in soil science must necessarily go hand in hand with the development of equivalent systems for disciplines closely related to soil science. The short-term objective must be to develop the use of these processing systems at the research and agricultural planning level (master plans and agricultural development projects).

In future, efforts should be made to solve the important problem of *interfaces* between the various systems invented so far. To overcome this problem, the soil scientists concerned will have to collaborate with the data processors, who will be responsible for solving data transcoding problems. Further guidelines on this point could be obtained from work already carried out, e.g. by Bie *et al*. (1975) for geological data.

Finally, the working principles

to be observed and the resulting limitations must be taken into consideration by all those concerned, in order to standardize the 'systems' internationally.

References

Bie, S.W., T. de Heer, C.J. van Staalduinen, J.R.E. Lieftinck & W.M. Dam (1975). Automated data capture from poorly structured geological borelogs. Geol. Jb. (Hannover) A 28.

Girard, M.C. (1975). Rapport de synthèse sur les objectifs et les structures juridiques et techniques d'une banque de données de sols. Publication Informatique et Biosphère. Paris.

Jongen, P. (1975). The use of data processing methods in land development studies. International Symposium on the 'World Food Problem'. Khartoum.

Riquier, J. (1972). A mathematical model for calculation of agricultural productivity in terms of parameters of soil and climate. FAO. Rome.

Ruellan, A. (1975). Rapport d'activités et programmes du Laboratoire de recherches de sciences du sol. Centre de Recherches. Rennes.

Sogreah (1972). Geotri. Technical note. Grenoble.

Van den Driessche, R. (1974). La banque de données pédologiques de l'ORSTOM. Cah. ORSTOM, série Pédol. *12.* 125-132.

Philosophies of computer-based data base management systems

K.G. Jeffery, Institute of Geological Sciences, Exhibition Rd., London SW7, England

Compter-based data management systems are a relatively recent phenomenon. As early as 1963, Honeywell produced IDS (Integrated Data Store) but the subject did not figure prominently until the late nineteen-sixties and early seventies. There has been an attempt to introduce a standard for data management systems (the Codasyl Committee, in 1971) but to date only one major system (DMS 1100, Univac) follows these recommendations.

The subject is in flux. However, some main themes are beginning to emerge and it is with these that we shall be concerned.

Two main themes to be considered here are data representation and the user interface.

Data representation

The word 'data' normally refers to a collection of values, the stored representations of information. Each item is referred to (explicitly or implicitly) by an item name. For example, the item name COLOUR may refer to a particular value RED in one instance. Similarly an item name DEPTH may refer to a value 9.36 (metres).

Now let us consider item names and data values from a programming point of view. The simplest relationship is direct assignment of a value to an item name (or variable name in most programming languages). For example in FORTRAN:

```
DATA COLOUR/'RED'/,DEPTH/9.36/
```

or

```
DATA RED/'RED'/
COLOUR = RED
DEPTH = 9.36
```

A program constructed with such assignment of values to item names is inflexible because it cannot accept a different value for the item name without reprogramming. This fact was realized very early in computing development and input/output statements were defined for the common languages to allow different values to be assigned to the same item name at different times. For example the values RED 9.36 could be read in FORTRAN by

```
READ (INCARD, 11) COLOUR, DEPTH
11 FORMAT (A3, F4.2)
```

and the same piece of software could accept the values

```
YEL5.42
```

However, the technique is restrictive to data type and format. This problem is overcome in all generalized data management systems by using a data designator that describes the data to the system and will accommodate almost any type and format of data values input. Each system tends to have its own style of data description, although the Codasyl Committee have recommended a standard Data Description Language (DDL), which is essentially an extension of the style used in the Data Division of a COBOL program. The use of a data description in systems design usually qualifies the system as 'data-independent'.

However, there is still a likelihood that a change of data may make programs processing the data unusable. Such a change is related to data structure. Let us consider an example with a data description (much simplified) of the form:

BOREHOLE HEADER INFORMATION
  1-99 STRATIGRAPHIC UNIT INFORMATION
    1-9 SAMPLE INFORMATION

indicating that for each borehole there is one set of information on the borehole itself, up to 99 sets of information on units in that borehole and up to 9 sets of information on samples from each unit. This is a fairly simple hierarchy.

The conventional software structure to match this would have loops to read 99 stratigraphic unit records after each borehole header record, and to read 9 sample records after each unit record. In FORTRAN the form might be:

```
      DO 100 I=1,100
      READ (INCARD,11) BORHOL,
      DATE, LATUDE, LONUDE
   11 FORMAT (A4,I6,2F10.4)
      DO 200 J=1,99
      READ (INCARD,12) UNIT,
      COLOUR, GRANSZ
   12 FORMAT (A4,A4,F7.3)
      DO 300 K=1,9
      READ (INCARD,13) SAMPLE,
      SIO2,TIO2, AL2O3 . . . .
   13 FORMAT (A4,10F7.3)
  300 CONTINUE
  200 CONTINUE
  100 CONTINUE
```

which assumes that there are 99 unit cards per borehole and 9 sample cards per unit. It is obvious that a variation in this structure would cause reprogramming, to recognize a 'sentinel' value and so exit from a loop or to have the number of succeeding records of the next lowest hierarchic level stored as a variable on the last card of the preceding hierarchic level.

If the decision were taken to record two types of sample for each unit, the structural changes to the program (and any other program handling this data) would be significant. Few data management systems are more independent of data structure than necessary to deal with the example given.

It is possible to divide systems into those that are independent of data content, those that are also independent of data structure and finally systems dynamically independent of content and structure. This variation in level of independence is usually reflected in the data description, which communicates the user's view of the data to the system.

The level of data independence attained by a system is largely a function of the way a system handles the logical structure of data. A major philosophic theme affecting system design is the stored representation, and structure, of the data.

User interface

The 'face' that a system presents to a potential or actual user is crucial. Through this interface, the user has to convey his requirements to the system, and the system has to present the user with the data in an acceptable form.

The Codasyl Committee have defined a standard Data Manipulation Language (DML) that is effectively an extension of COBOL.

The DML required by any particular user depends on the use he intends to make of the system. At one end of the scale a user may wish to command the system to 'find all data with references to boreholes in Great Britain' or to follow that request with 'draw borehole logs for them'. This an extremely 'high-level' type of request where the user expects the system to interpret his requirements and carry them out. Such a user interface has a high power ratio; one unit of user thinking effort requires considerable computer effort to obtain a satisfactory response.

At the other end of the scale, a user with an interest in the structure of information may wish to control the storage, access and retrieval of data to the detail of the movement of the heads on a disk drive, or to the bit-level coding of data values.

Most data-management systems choose an interface that falls into one of two categories. The interface may be described as 'user-friendly' or 'programmer-friendly', where a user is a person who need have not a knowledge of programming. This concept is of major importance in system design.

Systems design philosophies

From the previous sections, we may generalize the situation. Data management philosophies may be represented by their position on two logical axes, each axis representing one of the aspects.

The data representation axis runs from data content description through to data content and structure description and probably in again to dynamic data-structuring and description; all with attendant philosophies of physical and logical data-storage.

The user face axis runs from the programmer level through to the non-programmer level and possibly further with increasing power ratios being attained between units of user thought and corresponding multiple units of computation.

Considering the 'user-face axis' first, we can group systems broadly into host-language systems and self-contained systems. Host-language systems have a programmer-user face and the data base is accessed by using extensions to existing programming languages. The Codasyl recommendations favour this approach, and several systems subscribe to this philosophy (e.g. DMS 1100). In contrast, self-contained systems have a user (non-programmer) face with easy commands, which are translated, interpreted or compiled into object code by the system to access the data base. Systems such as GIS (IBM) have this philosophy.

On the data representation axis, systems are found all along, and discrete clusters are hard to find. Most existing data base management systems fall between simple data content description and simple description of content and structure. However, the data description is only the interface between the user and the system for communication about the data. Behind this face, the system usually has another communication link, between the logical data content/structure as specified in the description and the physical representation of the data in magnetic store. Although the physical storage of data is not the direct concern of the user, it is of prime importance to the efficiency, flexibility and scope of the system.

With this criterion, systems cluster into two main groups. One group has a physical structure closely resembling the logical structure. Logical records consist of segments that may or may not be present, and may or may not be repeated at different hierarchic levels. This group of systems stores segments associated with the logical root segment together, or in a way which allows rapid linkage. This usually implies that access to a data item in one of the lower hierarchic segments may only be achieved by linkage through the root segment and all relevant segments between the root and the required segment. This philosophy is followed by *non-relational* systems.

In contrast, there is a small group of systems which represent data a different way. Assume that we observe only a subset of the real world, and that we record only a subset of the observed world. The recorded world is thus a representation of the real world. Items in the real world are related by characteristics. Items in the observed world are related by observable (and observed) characteristics. Therefore it should be possible to relate data from the collected world by characteristics.

Take a simple example: people. All people have hair a slightly different colour. Therefore each person data item has a particular place on an axis labelled 'hair colour'. Similarly each person has a different height, and so occupies a unique position on an axis labelled 'height'. Furthermore, each person has a unique position on axes labelled 'weight', 'eye-colour' and so on. With enough characteristics, one can uniquely define the position in n space (using characteristics as coordinate axes) of any one person (or data item). This concept in a computer-based system is referred to as *'relational data base'*.

The logical storage form in a relational data base reflects the relationship of data items in sets. A set may have n items (or no items and thus be a zero) and these are stored in adjacent logical positions in the data base. However, it is probable that each item is a member of more than one set, and that each set relationship. An item is thus logically located by the charcateristics which identify or describe it. Furthermore, access to that item may be achieved by logically following the set of which it is a member and whose characteristic is one of the retrieval criteria. This procedure is common in manual systems for following a column of data values to find the relevant item.

Access along a set (or vector of storage positions) for each characteristic is a rapid method of reaching a data item. However, simultaneous access along two sets would significantly decrease the search time. This may be achieved by storing logical arrays or matrices of values. Again, the there is an analogy in manual systems. The common storage of scientific information in tables of results is to allow rapid access to any data item by two of its elements.

With increasing numbers of simultaneous accesses, the retrieval speed for a data item increases. However, most computers still process sets sequentially through the central processor so the benefits of n-directional simultaneous access are not yet realized.

This relational philosophy has been described extensively (by E. Codd of IBM) but few existing systems subscribe to this philosophy. Mark IV exhibits some of the concepts and G-EXEC is designed for relational processing.

The problems of accessing segments of lower hierarchic level in non-relational systems has been mentioned. In contrast, with a relational system, the characteristics of the data item(s) cause access to be rapid from any direction related to one or more characteristics. This best illustrated by an example. Taking the structure used before:

BOREHOLE HEADER INFORMATION
  1-99 STRATIGRAPHIC UNIT INFORMATION
    1-9 SAMPLE INFORMATION

let us assume that the data is about a chemical component, and that one data element is ARSENIC.

A scientist may require information about ARSENIC from all samples from all boreholes in an area. Using a non-relational system the software would probably access each borehole, each unit within borehole and each sample within a unit (minimum of three direct disk accesses) for each value of ARSENIC. In contrast, in a relational system, all data on ARSENIC would be stored in a logical vector of storage addresses representing that characteristic and so the only operation required would be a maximum of one direct disk access per value. With increasing complexity of data base structure, a relational system is more economic in access speed.

Implications of data base system philosophies for the user

We shall now explore some of the facets of systems from different philosophies. To obtain a useful perspective, it is best to consider the implications for the user, trying to utilize a data base in his work.

In a typical non-relational system with host-language interface, the user would have to code his query in an extended high-level language (such as COBOL) and could probably only obtain output in the form of a listing. The access method would probably use only the root-segment keys and so, for a novel query, the access time would probably be excessive. In a typical relational system with 'sophisticated' interface, the user would only code a succinct request and could obtain output in a variety of ways including graphic display and statistical analysis. Furthermore, the access to the data base would be more rapid.

The essential difference is the appreciation of the relation between user effort and computer effort. For a typical host-language system, the user effort is greater than in a self-contained system. The philosophy is carried further in an implicit manner. A host-language provides the basic elemental expressions to construct a program for statistical analysis or graphic display, whereas in a self-contained system such facilities must pre-exist as 'process-modules' to be called on command by the user. Finally, the non-relational system is geared to efficient production of data for similar users with similar queries and well defined access requirements. The data base is therefore usually predetermined and relatively static. In contrast, the relational data base is geared to providing rapid access for queries based on characteristics that exist in the collected world, and therefore presumably in the observed world,
which is the model for ideas in the user's mind. These queries follow no well defined paths and may entail different structural views of the same data.

For most scientific needs, a relational system with a high level user interface probably provides the most cost-effective solution.

Implications of data base system philosophies for the installation

Although the user's needs must take precedence in any systems design, it is advantageous to optimize installation efficiency as much as possible within the user-defined requirement. The effects of different philosophies on typical installations will now be considered.

Host-language systems require a compiler for the host-language plus data base expression extensions. It is usual for these extensions to require support in the operating system for file storage and access. Furthermore, it is usual for the implementations of host-language on different computers to differ in the response to a particular language statement. COBOL is particularly prone to this problem. However, there is the advantage that any user who can program in the host-language can (with minor extra instruction) access the data base rather than a personal file.

In contrast, self-contained systems interpret the retrieval commands and may or may not require a language compiler. If a compiler is required, the standard language (without extensions) is all that is expected. There is no demand on the operating system for more support than that required for simple personal files. The IBM GIS implementation, for example, requires only standard OS (operating system) files.

Although at first sight, host-language systems may seem more machine-independent, it rapidly becomes apparent that they tend to be less independent of operating system. Self-contained systems usually have a simple executive system within them to control the interface between the data base and the operating system of the installation. This executive system commonly provides run statistics and accounting information without special request. This feature is important because the future trends in system design are likely to involve system self-control and self-optimization. It will require system-generated statistics on file usage, access paths and software usage.

A view of the future

We are now in the position where more and more computer users are accepting the basic advantages of data base systems. However, many users, particularly in the commercial world, are softening the impact of data base technology by extending available interfaces (common computer languages) to encompass the new facilities.

A group of scientifically orientated users are demanding more of data base systems. They require lower user effort relative to computer effort. This requirement demands sophisticated interfaces requiring simple succinct commands. It demands a wide scope of processing capability in ready-programmed, well defined procedures which the user can call to provide him with output in the form he requires.

The exciting development is the relational view of data. It now seems possible that data may be stored in the computer in a form resembling and reflecting the natural relationships of the data items themselves. Computer storage will then resemble closely storage in the human mind (at least in logical structure) and so human generated ideas can quickly be matched in the computer system allowing rapid satisfaction of user demands for data on hypotheses being tested. At last, it seems possible for the user and computer to work together to solve problems and to generate ideas.

Data management: general systems and particular needs

T. Gordon, Geological Survey of Canada, 601 Booth St., Ottawa, Ont., Canada

Introduction

Organizations involved in earth science surveys and evaluation of earth resources face major problems in the field of data management. Large amounts of information obtained in the past, as well as growing amounts of new data, must be rapidly retrieved and processed to produce a wide variety of reports for scientific users, planners, and policy makers.

Computer technology has been called in to provide solutions to this problem. In particular, data management -- computer programs intended to replace myriads of clerks and filing cabinets -- are being evaluated to discover how they might aid in processing scientific data. There are hundreds of packages referred to as 'general data-base management systems' or GDBMS. They have been designed primarily for commercial uses -- for example payrolls, accounts, and inventories; and evaluation methods for these applications are well established (Kuss & Madill, 1972). Much less is known about the use of GDBMS for scientific applications.

In several years of examining such systems for the Geological Survey of Canada (Gordon & Martin, 1974a), I have discovered only one rule of thumb: general systems *cannot* be evaluated in isolation from the particular characteristics of the group they are intended to serve. A system considered ideal for one set of users, may well prove catastrophic for another group. Conversely, a system dismissed as 'useless' for some applications may provide efficient and economic solutions to other problems.

This paper is my analysis of the important characteristics of data management systems, organizations, and users of data; all of which must be considered in fitting general systems to particular needs. Although it is based on specific examples, I hope it contains generalities of use to others facing the task of system evaluation.

Definitions

'Generalized data-base management system' has been used to describe so many different software packages with such a range of capabilities and costs that the term has become nearly meaningless. In this paper 'file managemen' implies management of simply related data (e.g. a bibliographic file) with a small set of output requirements (e.g. almost all searches are to provide a reference list for literature research).

'Data-base management' implies management of complexly structured data (e.g. bore-hole data involving lithologies within time-stratigraphic units within a single well) with a wide variety of unrelated uses (e.g. retrievals for purposes of stratigraphic correlation and retrievals to determine petroleum recovery capabilities).

Although data-base management systems must handle complex data structures, both data-base management systems and file management systems must have the following three capabilities.

- Define/load - users' data, in machine-processable form, along with a description or definition of the file or data-base contents, must be processed into a form permitting future access by the system. At this stage the system should provide data checking or validation capabilities to intercept errors in the input information.
- Retrieval - selected data items must be retrieved, by some retrieval criterion or test of related data items. Boolean logic in the retrieval request and some output formating are essential.
- Update - it must be possible to delete, add, or change data items already in the file or data base.

Post-retrieval processing of data, for example statistical manipulation or map plotting, is not a common capability of data management systems. The requirements for such facilities should be evaluated separately from the considerations discussed in this paper.

Constraints

The process of selecting a data management system for a given set of applications is usually a process of elimination. Constraints, imposed by the organization and the users, combine to exclude many systems from any serious consideration. The constraints that I have found important are outlined below in the form of a crude questionnaire. An evaluator should find that answering these questions limits the number of systems requiring detailed study.

Organization constraints

1. Is the organization committed to a particular computer installation? If so, what manufacturer and model?

Many data management systems can only operate on equipment of a particular manufacturer. Of these, many require a minimum configuration of operating system, memory, and peripheral storage devices. Some systems, claimed to be 'portable', require extensive re-programming before they are adapted to a new hardware environment.

The Geological Survey of Canada, for example, is committed to use of the Computer Science Centre of the Department of Energy, Mines, and Resources (EMR). This installation is a Control Data Cyber 74. Although it has more than adequate memory and disc space to handle some very sophisticated data management systems, many systems specific to other manufacturers (e.g. MARK IV) cannot be implemented on it. Organizations committed to the exclusive use of computers with limited memory or an uncommon manufacturer will find their choice of data management system even more restricted.

2. How is computing paid for?

Some organizations face the choice of paying for rapid and powerful outside computer services or accepting slower, more cumbersome processing on a nearly 'free' in-house machine. The latter alternative should not always be scorned. Overnight turn-around and inelegant computer science may provide perfectly adequate data management services for many users.

For several years, EMR operated their own IBM 1130 providing slow but inexpensive data processing. Users insisting on more sophisticated computer power has to pay higher rates at commercial service bureaux.

3. What software does the installation maintain and how well is it supported?

All installations must have sufficient staff to maintain the particular operating system on their machine. Many are also prepared to support data management systems supplied by major manufacturers. Support for small special systems without major manufacturer backing is somewhat harder to obtain. The user organization may be forced to employ its own computer science expertise to maintain a 'home-made' or limited distribution system. This is not necessarily bad. A large group may have adequate computer staff, experienced in the operation of their own data management system. The enthusiasm and skill of such people can more than compensate for minor shortcomings of locally developed software. On the other hand, organizations with few scientific computer staff should be wary of systems that may require too much in-house maintenance.

The EMR Computer Science Centre, for example, supports the data management systems MARS VI and SYSTEM 2000. Support for systems such as SAFRAS (Sutterlin, 1975) or G-EXEC (Jeffrey & Gill, 1975) would generally have to come from user groups. The Geological Survey, one such group, would find it relatively expensive to maintain one of the latter systems (Gordon & Martin, 1974b).

The United States Geological Survey, however, has sufficient staff to support and develop their own system GRASP (Botbol & Bowen, 1975).

User constraints

1. Do the prospective system users wish to have 'hands-on' access to the data? If so, how much computer expertise do they possess?

Scientists engaged in research on data they have collected themselves generally wish to browse through their files and update and retrieve data items at will (Hutchison & Roddick, 1974). On the other hand, users of archived data (Whitmore, 1974) are usually content to submit a general retrieval request to a service person and receive a map or report without any concern as to how their request is fulfilled.

A systems to serve 'hands-on' users must then permit retrieval and update commands to be issued in a form compatible with the user's expertise. Some systems (MARS VI, for example) have easily learnt procedures that can be taught to inexperienced users in a few days. Other systems -- such as TOTAL -- require the hands-on user to be an experienced programmer.

2. How predictable are the retrievals and updates? That is, are the data to be used in a routine production scheme, or are the frequency and nature of access unknown?

In our Survey, one major type of data management problem involves processing of large amounts of numeric data to produce contour maps (Grasty, 1974; Holroyd, 1974). This type of application is often better served by custom-designed software than general data management systems. At the other extreme, some scientists require an extensive browse capability -- that is ad hoc retrievals and updates with very rapid turnaround (Hutchison & Roddick, 1974). Interactive retrieval and update features are thus essential for any data management system serving such users.

System characteristics

For a user, there are several characteristics of general systems themselves that require special attention (Gordon & Martin, 1974b). These can be grouped into three main categories:

1. Availability of maintenance and support.
A system that is well supported by its distributor, although expensive, may be worth the extra cost over an inexpensive system that requires the organization to divert its own resources into computer science.

2. Ease of use.
This broad category includes such things as availability of documentation, training requirements, and syntax of user language. These characteristics are best evaluated by the experiences of other users.

3. Technical characteristics/costs.
Systems must be examined to ensure that they will support the logical data structures required by the various applications. In addition to the operations of define/load, retrieval, and update, many systems offer report-generation and data-analysis facilities. These may or may not be worth the acquisition costs involved.

Operating costs can only be evaluated by experts familiar with the physical data organization of the file management system, the accounting algorithm of the installation, and the types of file access required by users. For example, some systems offer rapid retrieval at the price of high storage and update costs, while rapid, inexpensive updating in other systems is obtained at the price of slow, expensive retrieval capabilities. The advice of computer scientists is essential at this stage of an evaluation.

Examples

In the Geological Survey of Canada there are three broad categories of user/application requiring the support of computer-based data management.

The first of these are users with a high level of expertise in mathematical applications of computer science, data that is almost entirely numeric, and applications that are highly predictable (statistical manipulation or contouring of large data sets). These users have developed efficient custom-designed systems for their particular applications. They would probably benefit little from currently available commercial data management systems.

Another group are individual scientists with small files including both numeric and text data and generally with unpredictable retrieval requirements. The computer expertise of this group varies widely. These users have been encouraged to use the Control Data file management system MARS VI -- an easy-to-use system supporting a simple data structure. MARS has a limited data validation capability which is often useful for geological field data files.

A third type of data management application is for archival files used by six to twelve scientists with common requirements for reference material (e.g. location of mineral deposits). Although the multi-user nature of this type of data would suggest a data-base system rather than a file management system, in practise we have found it extremely difficult to agree on the structure and data content of such an archive. The results have been simple file structures with only few data items. Again, MARS VI is considered an adequate solution. As archival files grow, however, and applications of the data become more sophisticated, it is probable that more and more use will be made of the data-base management system SYSTEM 2000.

An application of this system has been described in Canadian Datasystems, February 1975. The Canadian Department of the Environment has established several large scientific data-base using SYSTEM 2000 at the EMR Computer Centre. Use of this sophisticated software permits Canada-wide access to a large amount of data, efficient transfer of data to applications programs, and control of the consistency of the data.

Conclusions

Evaluation of general data-management systems for earth science data involves consideration of organization constraints, characteristics of users, and uses of the data themselves. When these limitations are clearly identified, the computer science evaluation of various systems becomes a much more meaningful enterprise.

Acknowledgments

I thank my colleague, Gwynneth Martin, for her part in the work on which this note is based.

References

Botbol, J.M.; Bowen, R.W.; 1975. Geological retrieval and storage program (GRASP). *In* Hutchison (1975)

Gordon, Terry; Hutchison, W,W. (Editor); 1974. Computer use in projects of the Geological Survey of Canada. Geological Survey of Canada, Ottawa. (Paper 74-60)

Gordon, Terry; Martin, Gwynneth; 1974a. File management systems and geological field data. *In* Gordon & Hutchison (1974)

Gordon, Terry; Martin, Gwynneth; 1974b. Computer-based data management in the Geological Survey of Canada - a preliminary proposal. *In* Gordon & Hutchison (1974)

Grasty, R.L.; 1974. Computer processing of airborne gamma-ray spectrometry data. *In* Gordon & Hutchison (1974)

Holroyd, M.T.; 1974. The aeromagnetic data automatic mapping (ADAM) system. *In* Gordon & Hutchison (1974)

Hutchison, W.W.; 1975. Computer based systems for geological field data. Geological Survey of Canada, Ottawa. (Paper 74-63)

Hutchison, W.W.; Roddick, J.A.; 1974. Sub-area retrieval system (SARS) used on the Coast Mountains Project of the Geological Survey of Canada. *In* Gordon & Hutchison (1974)

Jeffrey, K.G.; Gill, Elizabeth M.; 1975. G-EXEC: a generalized FORTRAN system for data handling. *In* Hutchison (1975)

Kuss, J.K.; Madill. J.P.; 1972. A technique for evaluation of generalized data management systems. INFOR 10:311-319

Sutterlin, P.G.; 1975. U.W.O. SAFRAS system. *In* Hutchison (1975)

Whitmore, D.R.E.; 1974. Status of computerized files in the Mineral Data Bank, Geological Survey of Canada. *In* Gordon & Hutchison (1974)

The role of automated cartography in soil survey

C.G. Johnson, Soil Conservation Service, US Department of Agriculture, Federal Center Buildings, Hyattsville, Md., USA

Introduction

In the United States modern soil surveys are used for a great variety of purposes requiring soil maps that are at a scale meeting a particular need and show the required detail and the groupings of soils needed for a particular interpretation. At one extreme, soil surveys are used to advise farmers and ranchers on the best way to manage individual fields and to assist foresters in the management of individual pieces of woodland. They are used by developers and engineers to make first decisions on locating and designing housing developments and on locating schools, small industries, and highways. They may also be used to locate sources of gravel, possible sites for ponds, and possible sites for sanitary landfills and other waste disposal areas. On a larger scale, soil surveys are used in land use planning to determine whether a paticular area is prime farmland that should be saved for farming or whether it is better suited for recreation areas, town sites, or areas of industrial development. Soil surveys are used by tax assessors for the equitable assessment of real estate taxes; they may be used to predict the environmental impact of changes in land use or the effect of new management practices.

All these interpretations can be developed from the standard detailed soil survey. In the United States such soil surveys are usually made at a scale between 1:15 840 and 1:31 680; most modern soil surveys are now being made at a scale of 1:24 000. Detailed soil surveys usually show contrasting areas as small as 4 acres. Soil surveys meeting the present standards are available for about one-half of the United States, including most of the intensively used farmland and the rapidly developing areas around cities. An individual soil survey is usually made for individual counties or parishes. The average soil survey covers an area of 770 square miles (2 000 km^2). At the present time, the 1 600 soil scientists in the United States complete about 80 soil surveys per year.

Soil surveys in the United States are made as basic soil surveys in which individual kinds of soils, phases of soil series, are identified. Differences among these kinds of soils may be important for some uses but not for others. Therefore, each pusblished soil survey contains tables and narrative information that predict the behavior of these kinds of soils for many of the uses mentioned earlier.

The basic soil survey is primarily a document for the soil scientists and other professionals who have been trained to use it. Someone who is not a soil scientist needs soil maps that have been interpreted for him. In other words, he needs soil maps in which the soils in an area have been grouped into classes that show similar potential or similar limitation for a specific kind of use. Ideally, for any kind of use, such a map shows no more than three or four classes and the scale and the degree of generalization are adjusted to show the whole area under consideration and eliminate unnecessary detail. Often, information about the soil resource has to be related to other resource information, such as land use, transportation networks, ownership patterns, geology, and many other kinds of information. This is done most effectively by preparing overlays that show pieces of land that meet a specific combination of characteristics. Ideally, such interpretive maps are in color so that critical areas and the pattern of occurrence of such areas can be readily recognized.

Interpretive soil maps and interpretive maps combining information on the soil resource with other resource information have been prepared manually. Such maps are excessively expensive, especially if various source maps have to be converted to a common scale and if the interpretive requirements are complex.

Automated cartography

Automated cartographic systems, in conjunction with automated systems for soil survey interpretations and other interpretations, can cut the cost of preparing interpretive resource maps and make them available to a great variety of users. We also expect that automated cartographic systems will play an important role in efforts to protect the environment. Models of the movement of sediments, fertilizer elements, and other pollutants in the soil landscape are being developed. For their practical application, information about the soils must be available in computer-compatible form. The advanced mapping system (AMS) that is being prepared by the Soil Conservation Service of the US Department of Agriculture provides such a system. I shall describe AMS and use it as an illustration of the tasks performed by an automated system. Originally this system was envisioned primarily as an aid in drafting basic soil maps for publication. Changed procedures for preparing soil maps have largely eliminated this need, and the system is now intended primarily for preparing interpretive maps.

AMS is an automated cartographic application that will produce base, topographic, soil, and interpretive maps ready for black and white or multicolor lithographic processing.

Data for producing interpretive maps will be obtained from published soil surveys. The automatic scanner will digitize and incorporate the soil information into the data base. The features will be recognized both automatically and manually for identification within the data base. The data will then be edited for accuracy and stored for later interpretive use. In addition to interpretive maps, base and topographic maps will be manually digitized. The maps will be edited and photographically plotted at the desired scale.

The computer hardware of AMS is designed around four minicomputer subsystems to be located at a central facility. They are the scanning subsystem, identification subsystem, editing subsystem, and automatic drafting subsystem. Each subsystem will independently perform one or more operations. One remote filed system for digitizing base and topographic maps will be located at each of three field caryographic units. The cost for software and hardware, excluding the automatic drafting subsystem, is $1 400 000. The contract was awarded February 28, 1975 to Computervision Corporation. Delivery is scheduled for February 28, 1975, and the equipment is to be in full production in May 1976. The scanning subsystem will automatically scan, at the rate of 4 square inches per second (25 cm^2/s), an individual sheet of a published soil survey. An average soil survey area has 70 map sheets. A map sheet will consist of an aerial photograph and one, two, or three overlays containing alpha-numerics, cultural, and line information. The aerial photograph will not be used in the AMS scanning subsystem. The overlays will be film positives and be in one of three formats:

1. Alphanumerics, cultural, and line information on one overlay.
2. Soil lines and soil symbols on one overlay and planimetric detail and place names on a second overlay.
3. Soil lines on one overlay, planimetric detail on a second overlay, and all alphanumeric information on a third overlay.

Scanning of the alphanumerics in AMS requires manual preparation and recognition. We will eventually fully automate the alphanumeric recognition to reduce time and manpower requirements. As the map sheet is scanned, the alphanumerics are stored on disks in raster format, processed into vector format, and stored on magnetic tape for use in the identification subsystem.

The identification subsystem will use the magnetic tape created in the scanning subsystem as input data. Once the map sheets of an area or county are stored on disks, the data are ready for identification. For data on one overlay, significant manual identification must take place. For data on two and three overlays, many software features will be used to automatically identify soil lines, roads, and drainage networks. Discrepancies occurring in automatic identification will be stored and displayed for manual identification. Alphanumeric information is displayed sequentially on a 19-inch cathode ray tube (CRT). An operator will type the text into the keyboard as the character string is displayed. The keyed text will then be displayed on the CRT in the same location and angle as the displayed line text. When this step is completed, the next character string will appear on the CRT for identification. A second operation that the identification subsystem will perform is the production of interpretive maps. First, a list of the soil mapping symbols and soil interpretations will be entered into the subsystem for processing. The processing will then produce a magnetic tape for either the edit subsystem for editing or the automatic subsystem for photographic plotting.

The edit subsystem will correct errors in the data base or digitized data, produce a plot of the overlay or map, and prepare magnetic tapes for photographic plotting of base and typographic maps. A plotted map will be given to an editor who will check it manually against the original map. Errors will be noted on the plot, and the plot will then be sent back to the dit subsystem for corrections. These corrections are made via a 19-inch CRT or a coordinatograph table, depending on the accuracy required. Magnetic tapes from the remote filed systems also will be read in, plotted, edited, and a magnetic tape produced for the automatic drafting machine.

The automatic drafting subsystem is already at the central facility. The drafting table is capable of drawing with pen or photo light head. The resolution of this subsystem is ± 0.002 inches (± 0.05/mm) and accuracy is ± 0.005" (± 0.127/mm). This subsystem will produce all the alphanumerics and line work for base, topographic, soil and interpretive maps.

The scanner software will record coordinate data for lines, solid and broken, symbols, and underlined alphanumerics for each sheet of a published soil survey. It will recognize the placement and orientation of underlined character strings. The identification software will tag all alphanumeric, cultural, and line information in the data base and store it on different layers for ease of extraction at a later time. It will link soil symbols to line segments. Line segments will be linked together to form polygons to calculate area, to classify soils according to a list of soil mapping symbols with their interpretations, and to eliminate common boundaries when desired.

When interpretive maps are requested, special identification software features will be defined for each map, including selection of character fonts from four fonts; selection from as many as 64 special line types, which allows a line segment to accommodate two or more different features that describe the same line segment; request for area calculations; the scale of desired map; and number of color separations desired.

Editing software will provide the ability to add, delete, or move a feature, correct alphanumeric data, change the endpoint of a line, connect a line to another line, eliminate a crossover, change the line type, add or delete symbols, join feature parts that come from two different sections of a drawing, extract a section of a map by defining as many as 100 coordinate points, eliminate common lines between adjacent soil types that are the same, and automatically join data that come from two different sources.

AMS will produce interpretive maps for 70 soil surveys and 1 000 base and topographic maps per year. The daily throughput of the entire system is 24 soil survey publication sheets per 8-hour day at the central facility. In addition, one base map and one topographic map will be produced at the central facility and at each of the remote field sites per 8-hour day. The system will be able to produce graphic displays and engineering drawings from sketches. In addition it can be used as a computer for processing other computer programs. The system requires nine people to maintain the required work flow at the central facility.

To conclude, AMS is an integral part in automating the soil survey program. Its intent is to automate the cartographic phase for producing interpretive maps from a completed soil survey publication. AMS will be integrated with ADP files on soil survey interpretations to automate completely the preparation of interpretive soil maps. Digitized information from the AMS system will be combined with digitized cartographic data from local and state planning agencies to produce plans in which soils are only one of many considerations for decision makers.

Automation in cartography

L. van Zuijlen, Topographic Service of the Netherlands, Westvest 9, Delft, the Netherlands

Objectives of automation in cartography

The demand for maps, topographic and thematic, is increasing. One of the most complicated problems for mapping organizations is the time-consuming conventional method of production of the final cartographic sheets.

The main objectives for the introduction of automation in cartography are
- to speed up production of maps
- to reduce cost
- to allow production of special-purpose maps.

Cost is a tricky problem. Special-purpose maps are becoming increasingly important:

The traditional multipurpose map is being supplanted by the special-purpose map. The multipurpose map contains too much information for most users, but it is not detailed enough. Automated systems open the possibility of producing special-purpose maps on request, by sorting and combining separate data elements available in the data bank.

Steps in cartographic automation

In map production, I distinguish between three main operations:
- collecting data
- processing the data in the mapping procedure
- producing a graphical display of data.

In cartographic automation, we can express these steps more directly, namely:
- collection of data by digitizing of topographic and thematic maps, and by statistical methods for thematic maps
- computer processing and storage of data
- data presentation.

For each of these steps, we need a special hardware configuration, the instruments needed for the job. In each cartography system, the computer necessary for processing and storing data is the heart of the system. Data collection and the data presentation are the inputs and outputs of the processor.

The collection of data can be a statistical collection of data for instance for census mapping. It can also be a translation of analogue data into digital form with a digitizer. With that instrument, the information is made accessible to the computer. It is translated into digits. Points on maps are defined as pairs of rectangular coordinates x and y for the planimetric situation (and sometimes z for elevation). Lines are represented by strings of coordinates. Areas are defined by line polygons.

The cartographic digitizer is an instrument for determination of x and y coordinates of elements (lines or points) shown on maps. The coordinates can be recorded on paper tape or magnetic tape.

We can divide the cartographic digitizers into automatically and manually controlled types (Fig. 1).

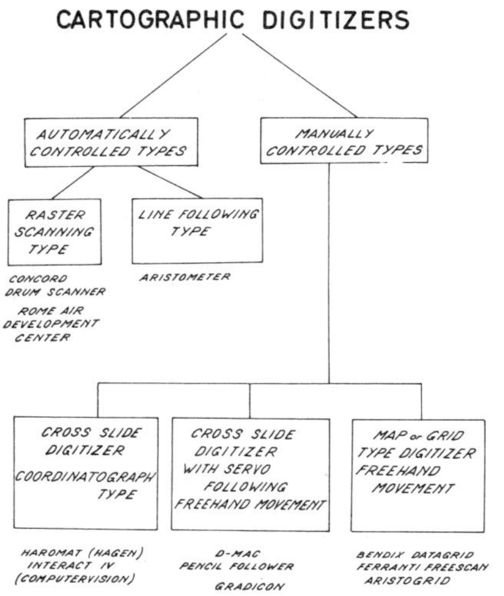

Fig. 1

Since manually controlled digitizing is time-consuming, attempts have been made to develop *automatically controlled equipment*. They can be put in two groups: the raster scanners and the line-following instruments. The raster scanners are mainly drum-type instruments with the map wrapped around the drum (Fig. 2). The scan head contains a light source and a photocell sensor, a photomultiplier, that is able to register the small dots (fractions of a millimetre) of points and lines on the scanned map. On raster scanners, the drum can rotate and the scan head is then in a fixed position only to move one interval after every run of the drum. The alternative is that the scan head moves parallel to the axis of the drum that makes one-interval steps only. The reflexions are recorded and compacted, e.g. the reflexions not representing a point of the map are neglected.

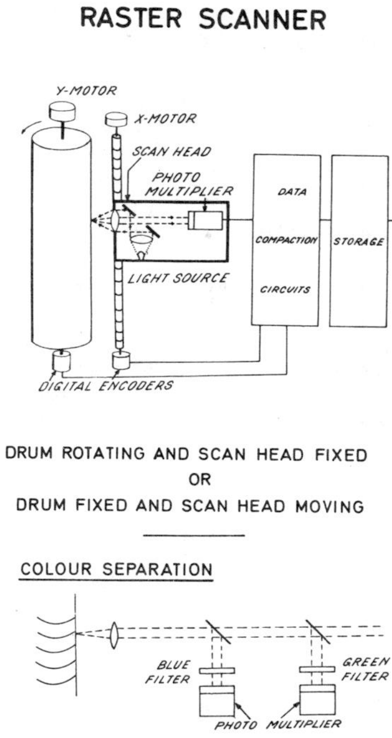

Fig. 2

The problem with this raster scanning is the coding of the features. The headers may be coded manually with an interactive cathode ray tube. Another option is to scan points and lines of one group of features from the colour separations of a map already made.

Some raster scanners have colour filters and can scan coloured originals.

The main part of an automatic line tracer is a tracking head that adjusts the direction traced. All positions are registered. The speed of such automatic line tracers is up to about 15 mm/s. However an operator is always necessary to assist the system, particularly at line junctions. The costs of these devices are so high that in cartographic practice manually controlled digitizers are normally used.

3 types of manual digitizers may be distinguished (Fig. 1):
- the cross-slide digitizer (coordinatograph type)
- the cross-slide type with servo following
- the mat or grid type digitizer

The first type (Fig. 3) is in fact a precision coordinatograph with a cursor in the place of the pricking device. One can put cross hairs of the cursor on a

CROSS-SLIDE DIGITIZER

(COORDINATOGRAPH TYPE)

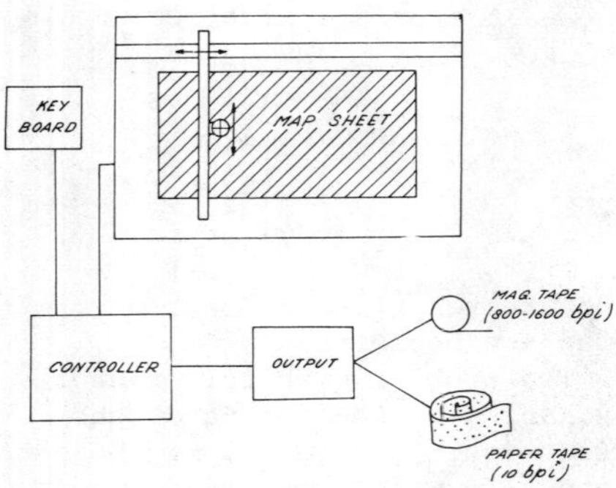

Fig. 3

point of a map and the position recorded on paper or magnetic tape. Lines can also be followed manually and the positions of the cursor recorded as a stream of coordinates.

A disadvantage of this type of digitizers is the friction of the beam that hampers the tracing. In cartography, the freehand cursors are preferred.

The cross-slide digitizers with servo have a cursor on top of the table generating a magnetic field that makes the servo instrument follow exactly the position of the cursor. The movement of that servo is converted into digital output by rotary encoders (Fig. 4).

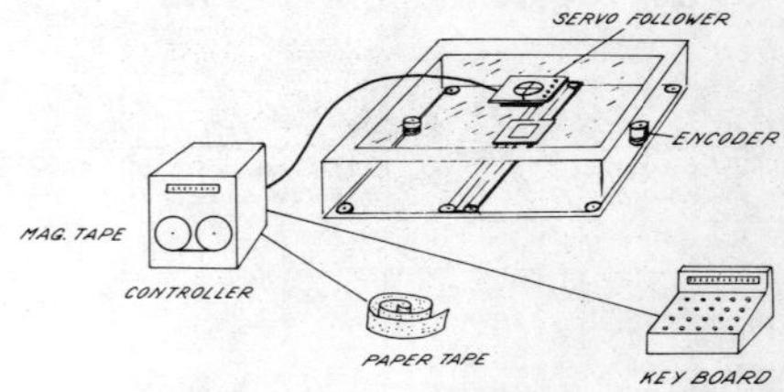

CROSS-SLIDE DIGITIZER

WITH SERVO FOLLOWING

Fig. 4

The mat or grid type digitizers (Fig. 5) use a 'digitizing tablet' and have only the cursor as the moving part. The 'digitizing tablet' is a special system of electrical conductors embedded in plastic. The induction voltage from the cursor into the 'grid system' of the electrical wires

MAT or GRID TYPE DIGITIZER

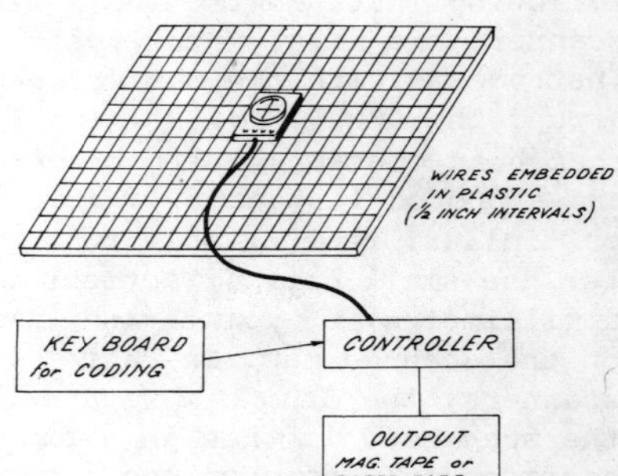

Fig. 5

creates pulses separate for the x and y axes. The resolution of the high-accuracy devices is 0.025 mm. The accuracy is about 0.15 mm.

However, accuracy depends, of course, completely on the accuracy of the fixed embedded wire-grid system.

Digitizing modes

On the cursor of many of the digitizers, three or four buttons are mounted. These buttons can, for instance, have the following functions:

a. recording of *x* and *y* coordinates
b. new line - same code
c. new line - new code

The button a, when pushed, initiates the recording of coordinates of the central point of the cursor (the cross hairs) on the tape after coding of the symbol concerned. It is important to reduce the work with the digitizer as much as possible.

On the controller (electronic cabinet) of the digitizer the operator can choose three modes for digitizing:

- point mode
- time mode
- incremental mode

As shown in Figure 6 with the point mode, the operator can

DIGITIZING MODES

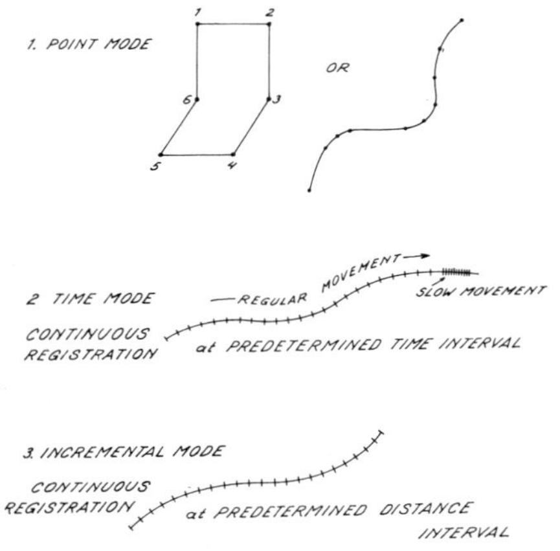

Fig. 6

choose points only. He puts the cross hairs on a point he wants to digitize. Having placed the cross hairs on the exact point, he pushes the recording button and the point is indicated on the output tape.

When the time mode is used, the operator can move the cross hairs along a line and when he keeps the recording button in the on-position, recording is continuous with a determined *time* interval. With this mode, the recording interval depends only on time. When the operator keeps the cursor in the same position, recording continues.

The third mode is the incremental mode. There too a string of coordinates is produced as the cursor moves along a line and the recording button is pushed. But recording depends on the *distance* that has been traced by the cursor. This distance-interval or increment is predetermined. In this mode, nothing is recorded if the cursor is kept in the same position.

For cartographic digitizing, the operators mainly prefer the point and the time mode. For large-scale maps only the point mode is used very often and for medium and small scale maps a combination of the point and the time mode.

Code for map elements

The map elements must be coded and recorded. On the top of the table of the digitizer, a keyboard can be used for this. The coding very often is done in letters, e.g. 'house' = HS.

An alternative is the menu technique. Part of the usable area of the tabletop (Fig. 7) is covered by a set of symbols in the map. The operator moves the cursor towards the menu and chooses the symbol he wants to digitize. He pushes the button

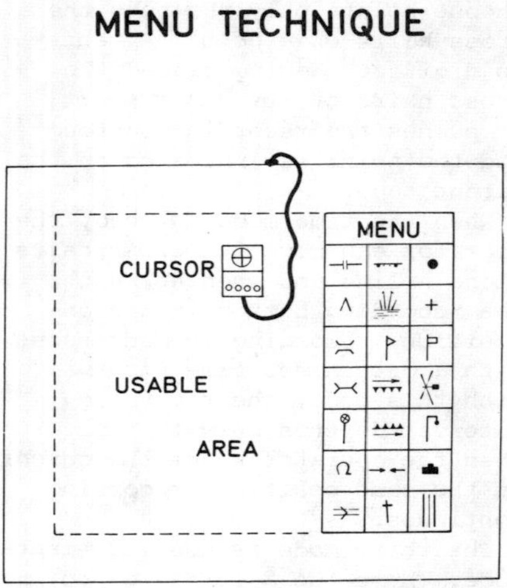

Fig. 7

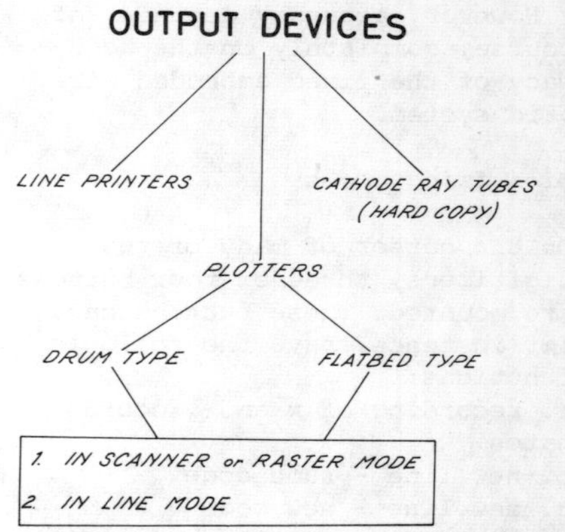

Fig. 8

and then procedes as for keyboard coding. The advantage of this method is that the operator does not need to memorize the codings and he sees the symbols directly displayed in graphical form. The disadvantages are that more memory capacity is needed in the controller and that a part of the usable area is not available for normal digitizing tasks.

A digitizing station consists normally of a digitizing table, the controller (a cabinet for the control electronics), a keyboard, a teletype and output devices (magnetic tape station and or a paper tape unit).

Output devices

Various devices exist for output for automated cartography (Fig. 8).

For thematic mapping, a line printer is often used for images working with area symbols. A few of the best known systems working with the line printer are SYMAP, LINMAP, COLMAP and AUTOMAP. Other types of plotters and cathode ray tubes (CRT) can use the raster-scanner mode or the line mode.

With the raster mode (mainly applied with photoheads as plotting tools), the light spot is moved in a series of straight lines, parallel to each other. Exposures in each of the scans form together a line on photo-sensitive material or on a CRT.

When the line mode is applied, a drafting tool of any type -- pencil, pen, needle, ball pen, photo head -- can be driven along a line determined by a string of coordinates. The result is drawn on paper, scribe coat, film or screen.

The drafting surfaces of a plotter can be on a drum or on a flat bed. Fore more accurate work, the flat-bed type is preferred. The drum plotter is often used as a check plotter. They usually have a higher speed of drafting. The drafting speed of a plotter is determined by the accelleration (time needed to come from standing position to the velocity wanted) and is important for production reasons. Work of higher quality requires less than maximum speed. The flat-bed line-plotters have an x and y beam system driven by motors. The accuracy of flat-bed drawing tables of a good quality can be 0.005 mm.

A special type of plotter, applying electromagnetic principles, is the Xynetics drafting table. The drawing head is controlled by magnetic fields. The drawing head hangs on a magnetic ceiling and an air cushion allows quick movement. The Xynetics plotter cannot do all cartographic work but with its high speed it is very suited for checkplots.

A similar high-speed plotter is the Calcomp 7000. Both instruments can draw with a maximum speed of 1 m/s.

Drawing tools

The drafting head of a plotter houses pens or scribing tools.

Some machines have two-position or four-position tool holders. Several pens or scribing needles can then be mounted, and the drawing and scribing in different line widths can be executed directly after another, under computer control.

Cutting knives can prepare peel coat masks.

Another drafting tool is the photohead or light-spot projector. It operates with a vertical bundle of light rays onto sensitive film. The drafting machine must stand in a darkroom.

A photohead may contain a disc with symbols, and characters that can be flashed onto the film thus reducing plotting. Such photoheads produce high-quality lines.

Data structure

Users of automated cartography systems have come to realize that the structure chosen for data representation in the computer is of paramount importance. In cartographic work the amount of data is very large but the data structure is usually simple. It must be possible to find point or lines in the data bank in order to edit (amend digitized data) or revise (update data) of a map.

A road limit on a topographic map may be coded RD followed by a string of coordinates (determing this road limit). A point on this road can be found by excluding all features not having code RD. This limits the search through the coordinates.

In cartographic work lines often have two functions. In smaller-scale mapping, a road limit could also be a ditch. Then the double coding RD DI would be needed.

Double digitizing can be avoided when a line section has a double function. An example is shown in Figure 9.

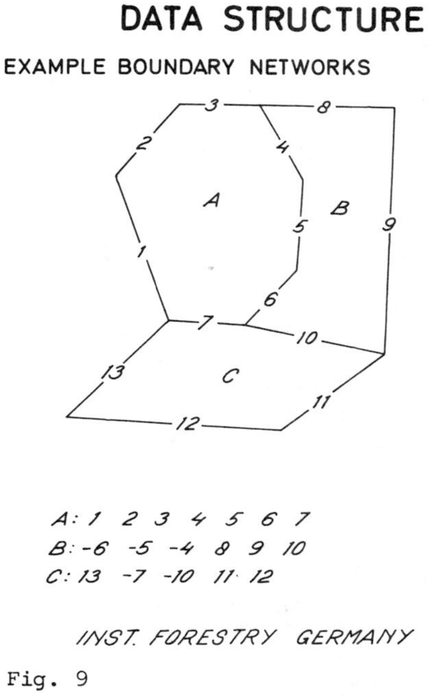

Fig. 9

For thematic mapping, polygons form important boundaries to delimit areas of type elements. In the Bundesforschungsanstalt für Landeskunde und Raumordnung in Bonn - Bad Godesberg, the procedure is to divide the boundary of every polygon into line segments.

On Fig. 9, area A is delimited by Segments 1-7 and these segments are digitized clockwise. Area B is limited by Segments 6,5,4,8,9 and 10 but now the same segments 6,5 and 4 are used anti-clockwise compared with area A. This is annotated as -6,-5 and -4. Thus redigitizing of the segments 6,5 and 4 is avoided. The same is done for area C, etc.

The data structure must permit points and lines to be rapidly traced, to shorten the time required for editing and revision. Data structure is also important for storage of data on magnetic tape, discs and other data carriers.

Cathode ray tube

The common television set is a cathode ray tube. In a vacuum glass container electrons hit a screen. The screen is covered by a phosphorescent layer. The spot which is hit looses electrons, becomes positively charged and emits light. The speed of electron movement is extremely high and the whole area can be scanned in a small fraction of a second. The picture disappears after a short period and to allow us to watch is has to be generated many (e.g. 60) times per second. This kind is called a refresh tube.

The other type of CRT is the storage tube. Here the picture has to be generated only once and it is displayed for at least an hour. The storage tube has 'flood guns' emitting electrons constantly over the whole screen, which remains bright.

The refresh tube needs more storage capacity because of the regeneration of the picture 60 times per second for instance. For most cartographic work, the storage tube is preferred because of the large amount of data to be plotted. With large data sets, refresh tubes may flicker as the system is unable to generate the whole picture in one cycle.

Cathode ray tubes can use raster and line mode.

Passive and interactive displays

On a passive display, a picture cannot be changed by the operator. On interactive screens, the operator may modify the picture and at the same time to change the data base from which the picture on the screen is formed. So with the use of a light pen or similar tracer one can indicate elements, change their positions or remove them. This depends on the software available. It may be necessary to display the entire map area on the screen. Subroutines should be available to enlarge a small section for detailed inspection. Other subroutines erase or add points or lines.

The size of the first-generation screens was small, typically 28 cm (11 in) diagonally. Now Tektronix produces storage tubes of 48 cm (19 in) diagonally (No 4014, if required with a device to give a paper copy of the screen image).

A new development of the interactive screen is the HRD - 1 Laser Display/Plotter designed by Laser Scan Ltd, Cambridge, England. It has a very large screen of 100 X 70 cm, a storage tube type with 5000 X 3500 lines resolution. Hardware allows one to draw pecked and chained lines, for character scaling and rotation. A cursor symbol can be positioned on the screen using a rolling ball.

The HRD - 1 Laser Display/Plotter can also plot. It produces a diazofilm copy.

Problems

Many problems still have to be solved in automation in cartography. One of them is generalization.

One can already think ahead during digitizing and choose and indicate points that are needed on smaller-scale maps as well (Fig. 10).

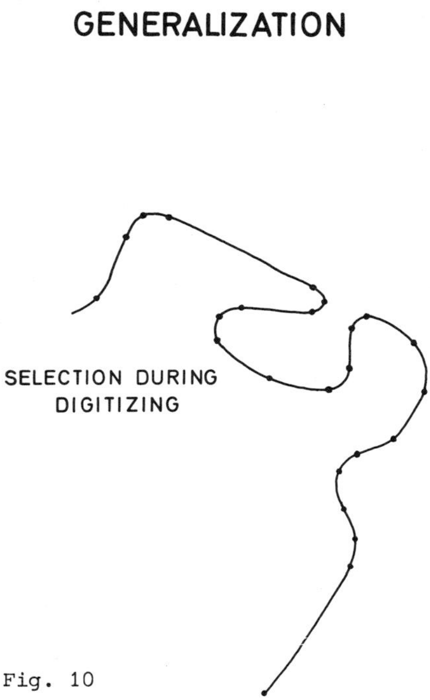

Fig. 10

Another important problem in topographic mapping is exaggeration (Fig. 11). With square streets one can remove the houses perpendicular to the street axis but if the street connections are at sharp angles this solution is not sufficient. Then an interactive solution may solve the problem. Computervision Corp. produces a system where groups of elements can be moved in chosen directions and over predetermined distance.

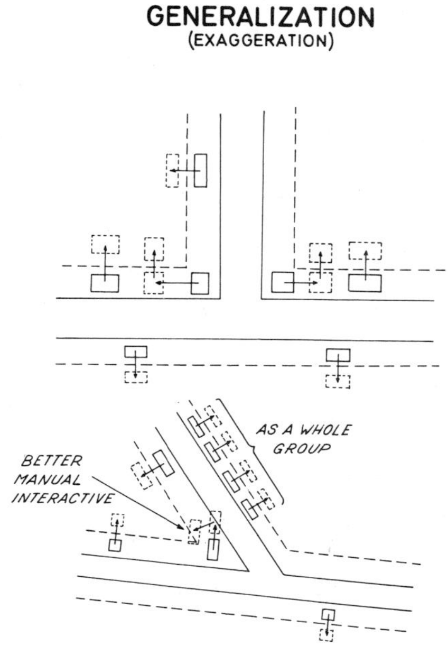

Fig. 11

The future

The storage of data for automated cartography is a major problem. The development of film-based video-disc, on which 2 X 10^{10} bits can be stored, could be an important step forward. Low-priced play-back devices for this disc can be expected. This opens broad prospects also for computer-graphics.

A wider use of laser in cartography, e.g. with raster scanners/ plotters (spot sizes between 10 and 1 μm), can be expected. CRT-lightheads for plotters will come. This hardware device will speed up the plotting of symbols.

The use of operational interactive systems also seem an important development that many cartographic institutes will want in the near future.

Nowadays, technical developments are so fast that one can hardly build up a philosophy for future developments. Within half a year, a chosen configuration could be out of date. However, if we compare the curves shown in Figure 12, we may expect that the application of automation in carto-

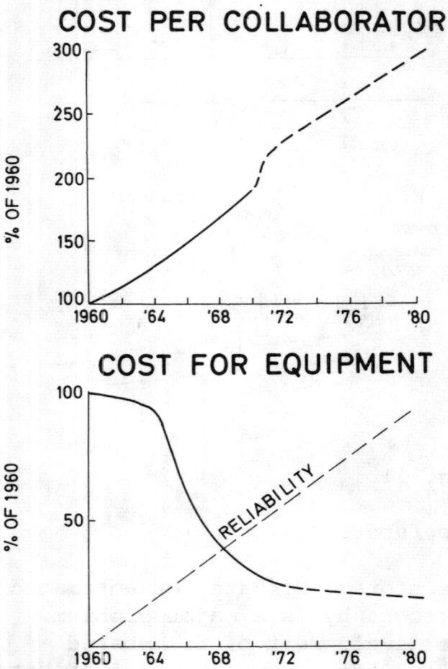

Fig. 12

graphy will rapidly increase in topographic and in thematic mapping.

But the introduction and development of automation in cartography in an organization will only succeed where there is a good teamwork and coordination.

The International Cartographic Association, through the work of its Commission III, attempts to encourage the dissemination of knowledge on automated cartography, and I am pleased that soil scientists are contributing towards this end.

Impact of automated data handling on data collection and manipulation

D. Teaci, Soil Science Institute, Bd. Marasti 61, Bucharest, Rumania

Automatic processing of data, in all fields of activity, requires an agreed national or international system of criteria and quantities in order to establish a translatable numeric language that may be handled by any group. Processing should encourage workers to obtain more complete data, should synthesize the data and bring research closer to the field problem.

Soil research, as a science and as a means of learning about land resources, covers a vast field. It relates to other natural sciences, with agriculture and forestry as well as with land use. Pedology cannot serve its purpose without clear concepts that are tested in practice.

Developed over a century in different schools, soil science has undergone several phases, from the geological to the naturalistic geographical. The present phase is a synthesis, as yet insufficiently uniform. Pedology is acquiring a self-discipline and is establishing itself as a modern and independent science.

However, when faced with the computer, pedology loses its former abstract character, when a master could state that 'this soil is undoubtedly chernozem because it is like the soils I saw near Kursk or in the Danube Plain', only to be opposed by a colleague who would assert that 'this cannot be; the soil is surely a prairie soil, like the ones I have seen in the Argentine Pampas or in Central Iowa'. No science can build and impose its concepts in this way, on the basis of a handful of masters and their pupils. Agreement will come only when sufficient quantitative and qualitative data on morphology, physics, chemistry, biology allow one to measure and record a soil's characteristics and correlate them with data on other soils. Computers subject scientists to close examination. Exact sciences, including soil science, grow or, if they do not serve their purpose and their clientele, they vanish.

In the past, soils were a matter almost reserved to farmers. Soil science was late to appear as a distinct science, while farmers continued to believe that the soil was their own business and that they knew enough about soils to use them properly. However, the impact of chemistry and machinery on nature forced farmers to change their opinions and to seek the assistance of soil scientists.

Informatics is now an essential aid and has had impact on the collecting and handling of pedological data, which serve, inter alia, for projects, decisions on land management, and agricultural economics. Why do I say 'impact'? Because nearly always, simple things are far more pretentious than complicated ones. Informatics needs simple and clear elements and does not operate with erudite formulas but with figures obtained by measurements and computations representing various characteristics of soils and soil-forming factors.

Because of the information explosion, soil science -- like other sciences -- cannot now be controlled by conventional means.

The volume of pedological and agrochemical data has increased even more than the number of specialists. Modern methods and equipment provide a much larger amount of data.

Closer links between research and farming requires a classification and parametric interpretation of these data. The large investments in research, and the need for a quick use of results in design and production processes, require correct and well founded interpretations.

Impact of informatics is most obvious where research and production meet. This is where all the knowledge of scientific research and of farmers and sylviculturists should strive together towards higher yields and higher quality with less labour. Practical interpretations have been mainly qualitative assessment, rarely based on technical parameters. Soil studies were mainly concerned with themselves or with the natural frame to which they belonged. To define the relationships between soil science and agriculture, forestry, biology and ecology, one must determine the parameters. Often in pedological and even agrochemical research, soil characteristics have been correlated with climate, vegetation, hydrology and lithology. However, only occasionally had they been correlated with biological parameters of plants in order to define how the various soil-forming factors and features control yield per unit area.

To process soil science data by machine, one must solve two main problems:

- 'Establishment' of soil science as an independent science by elaboration of parameters as complete as possible for all characteristics of the soils.
- Determination of the relation of pedological and ecological parameters with economic and technological requirements of land use for agriculture, forestry, or other human activyties.

The use of strict parameters to define soils has been a main concern for almost 75 years. Several data recording systems have been developed, of which the project of FAO (1972-1975) appears as the most universal. Correct and complete determination of the entire range of soils parameters helping to determine and characterize a soil profile and the effective range of magnitude of the parameter's variations are prerequisite to any information system in soil science.

Pedological data should represent an area and not just a single profile or 'pedon', so an area should be completely and comprehensively characterized. Accordingly, pure pedological and data on the natural, environmental and perhaps human factors for a larger area should not be separated.

Self-discipline of soil science is a consequence of the impact of modern computer techniques, as these require clear unambiguous numeric answers. It is irrelevant whether attributes and values are codified or nor, but that they be correctly interpreted for practical purposes.

The needs of various users of soil science generate another field of impact. Evasive answers by experts are no longer satisfactory. Users need clear instruction on how to do the particular job, for instance land reclamation, fertilizer application, or choosing the best crops for an area.

Scientists were taken by surprise with such demands. Until recently, they were not concerned with the determination of production functions of each ecological factor individually and of the entire range of factors influencing an area. Anyway, a complete study of production functions would have been impossible without adequate electronic equipment and modern processing techniques. Therefore the needs of soil research would not have been met without a coordination of efforts to determine the production functions of all factors in the growth and yield of both natural and cultivated plants.

Research in this field is being carried out throughout the world. However, information is slow to circulate, so that research is often undertaken on solved problems.

To designate the activity of correlating data on soil, on soil-forming and human factors with yields of grain or fruit, we have coined the term ECOMETRY. However, ECOMETRY should not be considered as a preserve of soil science. It should involve specialists in various other fields, such as climatology, hydrogeology, hydrology, plant techniques, biology and biochemistry, biogeography?

A task of such magnitude requires both explorative research and long-term trials in natural and controlled environments.

An example of this concept is the system used now in Rumania for the ecological justification of agricultural zoning, using a prescribed method for collecting, coordinating, processing and issuing of data required to solve technical and economic questions specific for each of the country's major agricultural zones or for microzones within each zone. (Soil Science and Agrochemistry Research Institute, 1975)

Ecological documentation requires maps of pedoclimatic zoning, land evaluation and technological characterization of agricultural lands. The ecologically uniform territorial units are separated on these maps by the following factors:

Lithology	nature of parent material
Climate	thermic resources water resources other climatic factors affecting plant growth
Relief	absolute elevation slope exposure form of relief
Hydrology	ground water table seasonal variations in water table ground water mineralization
Soil	morphological features physical features chemical characteristics

There are 39 elements for the characterization of the above listed factors, each of them subdivided into several degrees of intensity or into values.

Such a wide range of the territory and soils' features requires dividing on the 1:50 000 scale maps of very small areas and the identification of thousands of 'ecologically uniform territories' in Rumania, each having its proper identification formula.

For correct and easy recording of the data, each elements of environmental factors or soils and their subdivisions has been codified.

Using the data on each uniform tract of land, land is evaluated for 25 crops and of other land uses.

Thus, primary interpretation of ecological data for production is achieved.

Simultaneously agricultural land is characterized technologically, the following aspects being investigated:
- the energy required for soil tillage;
- accessibility of land for mechanization;
- need of erosion protection and control works;
- land capability for irrigation;
- need for drainage;
- need for embankments;
- need of works to prevent or control salinization and alkalization;
- liming requirement;
- fertilizer requirements;
- technological features of current soil management operations;
- problems of soil pollution.

For these characterizations, the valves of the features for each uniform tract of land should be considered. Several characteristics may be combined according to present necessities of land use but determined by its intrinsic features. Investments and costs per hectare should be investigated by determining the reclamation and management technologies required.

The possible effects of land improvement are also emphasized and should be added to the initial production potential of the land, so as to express the improved production potential of each part of the territory -- an essential element in any land zoning, or forecasts of production.

This huge work is under way. It could never have been started without computers.

Reference

Soil Science and Agrochemistry Research Institute (1975): Instructiuni pentru elaborarea lucrarilor de raionare pedoclimatica, bonitarea si caracterizarea terenurilor agricole in vederea zonarii productiei agricole. (Directions for research on soil-climatic zones, land evaluation and technological characterization of agricultural lands). Bucharest.

Sampling, classification and quality control

R. Webster, Soil Survey of England and Wales, c/o ARC Weed Research Organisation, Yarnton, Oxford, England

Introduction

In his paper on data collection, Teaci (p.61) has made three important points:
1. In most instances, data are collected about soil so that they can be used for practical purposes. We therefore observe and record those properties of the soil and its environment that bear on our current or intended use of the soil.
2. To interpret information consistently and precisely for these purposes, we need information that is recorded consistently and precisely, and so we should *measure* the properties of interest, or surrogates of them, whenever possible.
3. We are most often concerned with *areas* of land, not just points nor profiles. Soil information systems have their biggest potential here, but also present the biggest problems for data collection.

Thus the aim of collecting data in a soil information system will usually be to enable us to provide reasonably accurate and precise practical statements or predictions about the soil of areas of land or about sites within such areas. There will be subsidiary uses, but we may regard them as by-products of the system. This broadly is the context of soil survey, and of this paper.

Quality of information

Automation enables us to treat very much more data than we could treat without, and by methods that would otherwise be impossible. But how ever large the computer a and how ever elaborate the management of the data, the value of an information system depends on the quality of the data available and their relevance to the use of the system.

When applying the term *quality* to data, we need to be clear about its meaning. Data recorded by novices in the field and from ill-equipped or inadequately staffed laboratories clearly cannot be relied on, and we may regard it as of poor quality. It is almost certainly not worth entering in an information system. Data from experienced surveyors and well-run laboratories will usually be reliable, and these are data that we shall usually wish to store. The data are of good quality in the sense that each individual value faithfully describes the condition of the soil at the site where it was recorded or from which the soil was taken. This is one meaning of *quality*.

In the context of soil survey, *quality* bears a further meaning. For in making statements about areas or types of aoil, we have to assume that the data obtained from point sources adequately represent the soil of the areas or types concerned. If the data represent the area well and allow useful prediction, then we can consider the data to be of *good quality*. If they do not, even though they are impeccable in the previous sense, then we cannot rely on them and we must consider them *poor quality*. In this sense quality is closely related to sampling and classification and is the sense in which I shall use the term in the remainder of this paper.

Sampling

Soil forms a fairly continuous mantle, and information about the soil of an area, at least in its intrinsic properties, necessarily derives from observations of a part -- that is, from a *sample*. We cannot record what the soil is like everywhere. Soil also varies from place to place. So sampling must be replicated to represent that variety properly and to provide measures of confidence of the data. When a property is measured, we usually want to know its mean, e.g. the mean lime content of the soil. When data are qualitative, we may wish to know the proportion of the area occupied by soil possessing a particular attribute or set of attributes. In either case, the precision of our estimates increases as the square root of the sample size. Thus, the more observations we make, the more confidence we can place on the values we calculate, and the better the quality of the survey. We should note, however, that generally many more observations are needed to estimate proportions than are required to estimate means with the same confidence limit.

Although replication increases precision, it does not ensure accuracy. We must also avoid bias to obtain good quality. Information from biased sampling is likely to be inaccurate and misleading. Bias is best avoided by choosing sites in such a way that every point in the area has an equal chance of being sampled.

Stratification

There are yet further factors to consider in a soil information system.

Prediction. Information derived from survey is often required for prediction, especially when the properties of interest are time-consuming or expensive to measure, or the area remote. Although our best estimate of the value of a property at a previously unrecorded site is the mean for the sample, the confidence that we can place on that prediction depends almost entirely on the variance of the property in that area. We can scarcely improve the precision, and hence quality, of prediction simply by increasing the sample size.

Regionalization. The soil in almost any large area varies so much from one part to another that average values for many properties have little practical value. It can be pointless, or even misleading to apply a mean value for the whole area to particular parts of it. In these circumstances the soil is not to be treated as a random variable. There is a spatial dependence, so that soil within a small region is more likely to be similar than is soil from more distant places. In the language of the Fontainebleau school of mathematical morphology, soil is a *regionalized variable* (Matheron, 1965).

This second feature of soil distribution can be, and often is, turned to advantage in survey. A soil is classified and regions of similar soil delimited within the total area. The area is thus *stratified*; the strata are the classes oe regions, and general statements about the soil are confined to each stratum separately. The information required is obtained from representative sites within each stratum.

In most soil surveys to date, the representative sites have been chosen intuitively by the surveyor concerned. When manpower is scarce and one can afford only one or two sampling points per class, choosing typical sites in this way is almost certainly best. However, the procedure relies heavily on surveyors' judgment to provide data of good quality. There is no way of communicating surveyors' confidence, and there is a risk of bias, especially strong when sites are chosen by inexperienced surveyors.

If there is enough labour, these deficiencies can be avoided by replicate sampling within classes and without bias. When this is done, each class can be described by its class mean with good sense and reasonable confidence.

There is a second advantage of class assignment, for the variation within each class will usually be less than in the area as a whole. This enables us to predict more confidence. If the decrease in variance is appreciable, and we do not recognize too many classes, we might achieve adequate confidence limits with a smaller total sample. In this way we can cut the cost of survey.

Quality of classification. There are, of course, many ways in which the soil of an area can be classified, and there are not all equally effective in increasing the accuracy of statements about parts of the area or in making predictions more precise. Within-class variance is not equally reduced. Nor is a classification for one soil property necessarily effective for all. That classification for which the within-class variance is least, confers greatest utility on our data, and is in this sense best. A classification in which the variance of a property within classes is no different from that in the whole area is useless for that property; there is nothing to be gained by classifying the soil that way. Thus the utility of information gained by survey depends not only on the quality of the sampling, but also on the quality of any classification applied to that information.

Although we may use a prior classification of an area to economize on sampling, we must beware of using the information so derived outside the context of that classification. It is not necessarily valid. Also, before deciding to make a prior classification, we should realize that it too requires effort. The effort can be large, and be unjustified by the benefits that accrue from it. When the benefits expected are small, then the resources required for classification are better transferred to increased sampling.

Numerical classification

In the circumstances described so far, I have assumed that classification, either before or after data collection, does not depend on the data that are stored in the system. Instead I envisaged a surveyor's dividing an area intuitively, or using systematic guidelines, mainly on the appearance and 'feel' of the soil and its associated landscape. The soil can, however, be classified on the data themselves, and the advent of computers has brought opportunity to do this quickly and consistently by simultaneous examination of multiple measurements in ways that are in some sense best. Other multivariate methods, such as component analysis and canonical variate analysis, enable relations among individuals and groups to be seen to best advantage, and enable us to study the process and effects of classification. We can also use the computer for optimal allocation of new individual profiles or sample sites to an existing classification.

Numerical classification is fascinating. The subject has expanded apace and we are still learning fast. There is space here only to mention those features and achievements of numerical classification most important to soil surveyors, and their value and shortcomings, as I see them, in a soil information system.

Nearly all the early methods of numerical classification created hierarchies, either by grouping individuals into small classes, and these in turn into bigger classes, and so on, or by successive division of the whole population or sample into smaller and smaller groups. If data have a hierarchical structure, the methods are admirable. If not, the resulting hierarchy is an artefact and the assignment of sample data can be seriously influenced by chance. No soil population studied so far has been shown to have a hierarchical structure, and few are clustered. Most seem to have remarkably even statistical distributions (e.g. Fig. 1) and for these a hierarchical representation is misleading.

Hierarchical methods of classification attempt to optimize the connexions between individuals. The minimum spanning tree, which is topologically equivalent to single-linkage agglomerative clustering, is a good example. The classes themselves are unlikely to be optimal. In non-hierarchical methods there is the expressed intention to optimize classification in the sense of creating classes within which variation is minimal. Thus they embody the same concept of good quality for a classification as used already. The methods involve creating a classification, evaluating some function of within-class variation as criterion of goodness, transferring individuals between groups, and evaluating the criterion afresh. The process is repeated until no further improvement seems possible. A true optimum can be assured if there are fewer than about twenty individuals (Edwards & Cavalli-Sforza, 1965), but if there are more, as

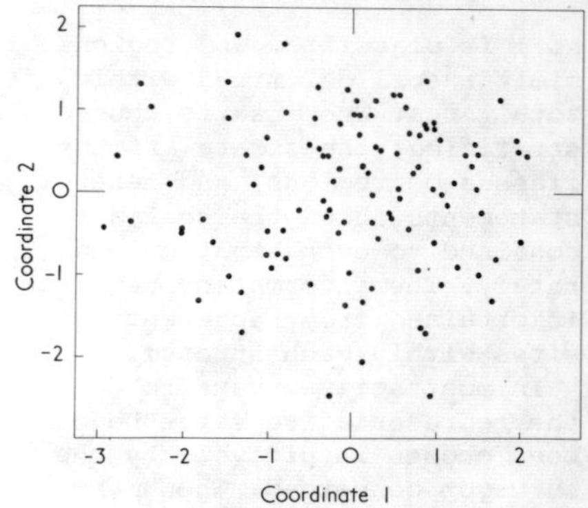

Fig. 1 Scatter of 111 sampling sites on the Ginniderra Experiment Station in the Australian Capital Territory projected onto the plane of best fit through morphological data (from Webster and Butler, 1976).

there usually are, then a local optimum must suffice. These methods are less elegant than hierarchical ones, and can be time-consuming even on the most recent computers. The only published uses of such methods in soil science are by the Netherlands Soil Survey Institute (Crommelin & de Gruijter, 1973; de Gruijter & Bie, 1975). Nevertheless they now seem much the more promising for soil.

Numerical classification has evolved as though the scientist little knows what criteria would be appropriate. It seems to me, however, that the pedologist usually knows some criteria in advance. He or she usually has some idea of the main groups that it would be profitable to recognize, e.g. brown earths and podzols; soils on shale, limestone and sandstone; soil that is good, bad and indifferent for rice production. In these circumstances it seems desirable to use this information. In non-hierarchical schemes, this is possible. The scientist can provide the initial classification, which the computer can then attempt to improve. Alternatively, the scientist may classify only a portion of his sites - say those that he finds easy to classify - transfer any that turn out to be misplaced, and allocate the remainder using similar rules to minimize the chosen test criterion. Webster & Burrough (1974) illustrate the advantages of this method.

Numerical classification, and the complementary techniques of vector analysis, clearly have merit. They provide good insight into the process of classification and enable us to discover structure in data. But classification of numerical data, however it is done, does not necessarily confer practical advantages. For even the best classification of quantitative data loses information that we have taken care and spent money to collect. At worst, it can completely lose the information that is most relevant to the real purpose of the survey, as Webster & Butler (1976) show. Thus having made measurements, we can best use them as they stand without generalization, matching or communication.

I believe that for practical purposes in soil survey, numerical classification needs further development, and in particular we should aim:

- to handle much larger numbers of profiles or sampling sites than now
- to embody the regional component of soil distribution
- to attempt to create reasonably stable classifications from sample data, i.e. classifications that remain valid when data from many more sites in the area are added.

The last is likely to present the biggest problems, for we then no longer have complete information about finite sets, but sparse information about infinite ones. Thus, I see the value of numerical classification for assigning sites for which sample data are available to regions and for general purposes, for example management of mixed agriculture and land-use planning. The recent work by de Gruijter & Bie (1975) seems a step in this direction.

Interpolation

Soil surveyors have devoted much of their research effort to classification so that better use can be made of data or field effort saved. However, if much expensive measurement is needed to use a classification or if classification hardly improves the utility of relevant data, we need to seek an alternative for worthwhile generalization and prediction. Interpolation is such an alternative, and we are only now beginning to appreciate its merit.

Technicalities apart, interpolation is simply a matter of inferring (predicting) values of variables at points from the measured values at surrounding points. This can be done in a number of ways, each with advantages in appropriate conditions. However, they do not all give the same results, and so, as with classification, we wish to know which is best: in which interpolated values can we be most confident.

Thanks to work by Krige (1966) in South Africa and Matheron (1965) and co-workers in France, we can now interpolate so that the variance is minimal and also obtain an estimate of that variance. The process has become known as *Kriging*. It has not yet been applied in soil survey.

The quality of interpolation depends not only on the method used, but also on sampling, especially its *intensity* (i.e. number of sample sites divided by area). In general, the more intense the sampling the better will be any interpolation from the data obtained. Sampling must, however, exceed some critical intensity for worthwhile interpolation. Consider Figure 2 in which variation in some soil property across the land is shown by the line AA´. If this line is sampled only at points u_1, u_2, . . . u_6, widely spaced as shown, then interpolation from them will quite fail to represent the wave pattern. The sampling interval is too wide. In this example, the sampling interval needs to be at least halved, to allow anything

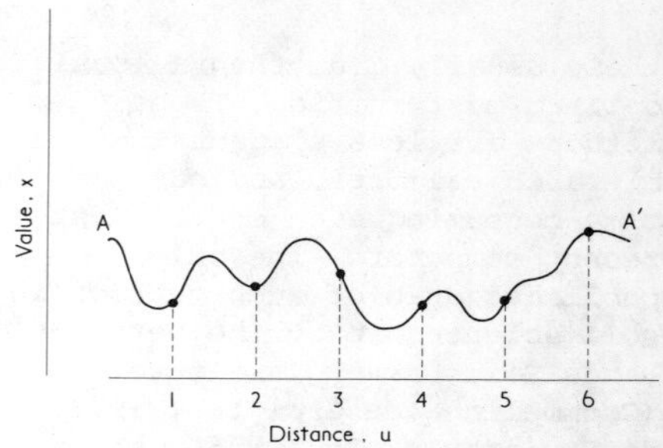

Fig. 2 A regionalized variable sampled at points u_1, u_2, . . . u_6 on a transect

of the pattern to be regenerated by the sample data. The *range* of spatial dependence in a soil property determines the minimum acceptable sampling intensity to interpolate that property. This minimum corresponds to a sampling interval of half the range. But since the range will not be everywhere the same, sampling must be intensified either throughout the area, or at least in those parts of the area that are already known to possess the more intricate patterns of variation.

Quality control

We are thus able to assess the quality of soil information, but can we *control* it?

Other things being equal, we know that:
1. the bigger our sample, the better our estimates of means and distribution functions
2. the more refined and relevant our classification, the better our predictions
3. the more intense our sampling, the better our interpolation.

However, we cannot say initially how good any of these estimates, predictions or interpolation will be, because we do not know the population variances or, for Case 3, the spatial covariances. Our first step in quality control must be to obtain estimates of these. In other words, a proportion of the resources for survey should be devoted to this initial part of the task. If the aim of survey is only to obtain a mean or distribution function (Case 1), much of the survey effort will be involved in this first stage. In the second case, we usually do not know what a classification can achieve until we have tried it. Nor can we easily exercise control in classification. For really important differences are usually obvious and agreed by all. When differences are less obvious, much depends on personal flair.

One of the more profitable tasks for the immediate future is to master the geostatistical techniques developed at Fontainebleau and elsewhere and apply them experimentally to soil survey. Experience gained should give good guidance in planning the main part of a soil survey.

However, even when we are armed with this preliminary information, we cannot usually design an economic sampling scheme because we do not know what precision to aim for. We can contrast our situation with that in manufacturing industry. There a customer requests goods of a certain quality from a manufacturer. The manufacturer, in order to win and retain that custom, must agree to produce the goods at a price the customer can afford. However, he will also wish to maximize his profit. So he will plan production so as to minimize his own costs while maintaining a quality of product sufficient for the customer plus a little for good will. In soil survey we should be able to do likewise; to minimize our sampling, and hence costs, to just sufficient precision for the purpose in hand. Unfortunately it seems that, with very few exceptions, neither users nor surveyors know what precision is adequate. Anf this is just as true of special-purpose surveys as of general-purpose ones, where such ignorance might reasonably be excused. Instead it is common practice to maximize precision is adequate. And this budgets and resources of manpower, or in some instances to pursue survey as a craft, almost regardless of cost.

Another task for the future is clearly to discover what precision is adequate for any given purpose. A possible way is to present the user first with crude information, and then to refine it until the user finds it acceptable.

Perhaps by the time this working group meets next we shall have some answers to these questions.

Summary

A system for handling soil information is potentially of most value for enabling reasonably sccurate and precise statements to be made about the soil over areas of land and for predicting soil conditions within them. The value of the system depends on the quality of the information it handles. Data of good quality derive from unbiased sampling with sufficient replication to provide precision in estimates of population parameters. Sampling must also be intensive if interpolation is intended. Prediction can be improved by classification, and the degree of improvement depends on the quality of the classification. Classification is not necessarily advantageous.

A preliminary part of any programme of data collection

should be to assess the variation present in the area of interest and to measure it. New techniques of geostatistics promise to be valuable for this. There is a need to discover what quality of data is sufficient for different purposes, and to design surveys accordingly.

References

Crommelin, R.D. and de Gruijter, J.J., 1973. Cluster analysis applied to mineralogical data from the coversand formation of the Netherlands. Soil Survey Papers No. 7, Soil Survey Institute, Wageningen.

De Gruijter, J.J. and Bie, S.W., 1975. A discrete approach to automated mapping of multivariate systems. Proc. Int. Cartograhic Ass., Comm. III, Enschede (in press).

Edwards, A.W.F. and Cavalli-Sforza, L.L., 1965. A Method for cluster analysis. Biometrics, 21, 362-375.

Krige, D.G., 1966. Two-dimensional weighted moving average trend surfaces for ore evaluation. Proc. symposium on mathematical statistics and computer applications in ore evaluation. Johannesburg, pp. 13-38.

Matheron, G., 1965. Les variables régionalisées et leur estimation. Masson et Cie, Paris.

Webster, R. and Burrough, P.A., 1974. Multiple discriminant analysis in soil survey. J. Soil Sci. 25, 120-134.

Webster, R. and Butler, B.E., 1976. Soil classification and survey studies at Ginninderra. Aust. J. Soil Res. 14, (in press).

Towards integrated soil information systems

S.W. Bie, Netherlands Soil Survey Institute, P.O. Box 98, Wageningen, the Netherlands

Introduction

Soil investigation agencies are the custodians of soil information collected in the past. They will also play a key role in the collection of future soil data. As in many other fields the introduction of computer-aided methods leaves such organizations in a dilemma: On the one hand they possesss archives of very large quantities of data collected and processed in the light of limited data handling facilities. This information makes up our current knowledge of the soil universe. On the other hand, many of the techniques and strategies used in manual data handling constitute sub-optimal use of the original data. Automated data handling has removed many previous constraints, and thus given us access to new and potentially valuable methodological approaches to soil investigation.

There is a paradox in the current development in soil information systems: automated data handling is initially being applied to existing data or data collected according to existing methodology. These are the very data that are likely to be least suited to automated treatment. But the strong points of computer-aided systems have not yet been translated into a practical alternative methodology. So some of the tasks at which computers are particularly powerful, are not being implemented in major systems now under construction.

This paper attempts to illuminate the paradox, and to suggest short-term and longer term responses by constructors of integrated soil information systems.

Inheritance reviewed

The principles of soil investigations have remained relatively unchanged over many years. Compatibility and continuity have been key words. The data collection methods aim at few observations that are 'typical' or 'representative'. Each site is fully described, often in a personal terminology, based on the principle of 'little, but good'. The naturalists' desire to understand the soil-landscape is reflected through the soil classification. Soil mapping is the art and craft of interpolating from the few known sites to the soil landscape as a whole. The map legend strives towards compatibility with other existing legends. The map image is generalised to efficiently convey soil information. For practical use the map thus created is interpreted and recoloured to display the suitability of the land for different uses.

Data compression and generalization have proved essential tools for cost-effectiveness when hard field data are few and far between.

Data compression and generalization have been seen as necessary tools when we have got much hard field data. Our small minds are unable to process large quantities of individual data. This is the situation faced by most large soil survey organizations to-day.

Modules of automation

The first attempts at automated handling of soil data came in the statistical manipulation of point information: the easily quantifiable attributes of the soil landscape and the associated laboratory analyses. The forest soil data of Scandinavia (Låg, 1962, (Troedsson, 1964), the intensively sampled soils of Belgium (de Leenheer *et al.*, 1968), the progressive commercial soil consultants in France (Jongen, 1975), the computer profile files of the US Soil Conservation Service (Flach, 1975) and their Canadian colleagues (Dumanski *et al.*, 1975): they all bear witness to the desire to have a more efficient archive that can also calculate means, percentages, and standard deviations. The introduction of computer-aided methods here has been uncontroversial. It is an upgrading of existing filing techniques.

About the same time, a numerical development began in soil taxonomy. Many experiments in numerical classification conclude that one can by computer allocate a soil individual to the same class as with manual classification techniques (e.g. Bidwell & Hole, 1964). This is also non-controversial to many. Only if numerical classes do not relate to the old classification does computer-aided classification give rise to debate.

It is not long since the first products of automated soil cartography appeared. The multicolour lineprinter maps of the US Soil Conservation Service (Nichols & Bartelli, 1974) represent a translation of existing soil maps into raster format, with subsequent recolouring for interpretation maps. It is a low-cost rapid alternative to the normal derived map. An early example of automation of the conventional high-quality soil map, is the drainage map of Abingdon, England, by the Soil Survey of England and Wales and the Experimental Cartography Unit (Soil Survey of England and Wales, 1973). The computer is here a pantograph, truthfully regenerating the lines and codes originally entered. The human hand has been replaced by the numerically controlled penholder, but the human mind that generates the map remains beyond automation.

Current moves towards integrated systems

Systems are currently being constructed around these modules. Automation consists mainly of the mechanizing of functions that so far have been carried out by hand. More efficient filing possibilities have given the opportunity to document the content of profile classes, both in terms of soil profile characteristics and their quality for interpretations for various land uses. We will shortly see such information used for the more consistent description of soil mapping units and on interpretation maps. Many links in the transfer of information will remain in the hands of the scientist. We have not yet learned to simulate decision-taking in classification, boundary drawing and interpretations in the computer. This to some extent overcomes tricky problems of format compatibility. Data structures of profile files, interpretation files and cartographic files are naturally so dissimilar that simultaneous retrieval from a number of such files in an integrated system are scarcely to oversee. Rigorous demands are being put on data base management systems used to organize such data. The *ad hoc* retrieval programs now applied by many reflect the lack of serious concern for integration. Whilst individual files may be easily handled by simple retrieval programs, such retrieval techniques usually break down when files reach production proportions, and almost certainly when attempts at integration of the models are implemented. Soil information systems appear to lag significantly behind current knowledge in information science.

Steps towards the integration of external data files (e.g. digital topographic files, meteorological files, land-use files, planning files) remain largely a dream. Again the lack of concern in data base management must take much of the blame.

A major difficulty in the coordination of modules in systems under development lies in the terminology. Human powers of association and interpretation are not easily transferred to the computer. Whatever solution is found, terminology will have to be standardized in the gathering of raw data. The French-speaking countries have taken a lead here (ORSTOM, 1969), and must wish them much success in their thankless task.

Many modern soil classifications are sufficiently logically defined that class allocation algorithms may be employed leading to automatic classification of soil indivisuals. Such allocation algorithms appear mainly to work on key principles, thus not relating to all previously classified individuals, only to the rules of classifying them.

Digitizing techniques, particularly developments in point mode digitizing with spline functions, and automated scanning, have greatly strengthened the autocarto module. The constant manual redrawing of manuscript maps to make a suitable copy for digitizing is a terrible waste of time. No doubt there is a simple technical solution to this problem in the world of modern materials.

The integration of modules that so far have developed, relies on major bridging efforts by the soil scientists. No doubt they can be made to operate. We shall surely see major systems, e.g. US Soil Conservation Service, Canada Department of Agriculture, Orstom in France, Netherlands Soil Survey Institute and others in routine production before the end of the decade.

But there are two major problems that remain unanswered in current development: economics and a suitable alternative methodology directed towards automated data handling.

We know very little about economics so far. In geology, geography and planning automations along conventional lines have proved expensive. Automated cartography in other fields have shown no great savings (e.g. topography, see Coles, 1974). The indications are that a much wider user of conventional soil data is required before economies of scale can be expected with this type of integrated system. Agressive marketing policies would need to support such a development. Whether our current methods of working, particularly for interpretations, are satisfactory to support such a marketing drive, remains in my mind an open question.

The second question has a more futuristic element: Does it make scientific sense to employ automated data handling techniques to support data arising from a methodology developed for manual data handling?

Integrated systems:an alternative approach

Current soil investigation methodology is - at least in part - a function of previous and existing manual data handling capabilities. Methods may be developed that are also able to exploit the capabilities of computer-aided data handling.

Computers are particularly good at handling large sets of information according to fixed rules (algorithms). There is no need to operate with subsets and generalizations at early stages in information flow. Only when presenting information to the human mind, need simplifications be introduced. That may be the very last stage of data processing. This contrasts with the early steps towards generalization in conventional processing.

It is then unnecessary to seek 'typical' observations, since such sites are only typical in relation to the generalised hypothesis applied.

For many interpretations, a pedological classification seem superfluous, involving information loss. Instead, a direct attribute-value approach, with calculation of covariance and similar indices for unmeasured attributes, could fulfill a similar role.

Utilitarians will therefore also argue the diminishing role of the general-purpose soil map, thus eliminating the need for human intuition often associated with the making of such maps. Special purpose maps may be generated directly; their boundaries generated from existing point data, possibly aided by existing line or area information. The classification here takes the very last stage, forming the map image through the legend. Many general-purpose maps could also conceivably be generated from similar information.

The alternative strategy is thus: pedological experience can be expressed in algorithms. Human effort is transferred from repeated manual application of experience to data, to the formation of once and for all machine algorithms, where the machine does the rest. This is precisely when computers become economic, replacing expensive scientist time.

This alternative approach to integrated systems has two consequences on personnel:
1. it releases highly qualified personnel from routine tasks where considerable experience nevertheless is necessary
2. it gives the opportunity to build an efficient production system around inexpensive lower technical personnel. In developing countries this could be the only feasible approach.

It cannot be disputed that there is a shortage of good algorithms to put an alternative system into production in 1975. In classification for interpretation, and in mapping algorithms, much work remains. The basic sampling design of the soil landscape is still an open question.

Towards which integrated system?

The choice of the type of integrated system is essentially a choice between different methodologies. The conventional integrated system utilizes the information after processing (data compression and generalization) by the human mind. The alternative integrated system maintains direct contact with the original data to the last moment. Mathematical algorithms simulate classification and mapping functions normally carried out by the scientist.

The conventional system has as its building stones almost 100 years of experience and tradition in soil science. It automates the routine functions of existing ways of handling data. It is natural for large soil survey organizations to choose this road towards automated soil information systems. There may be administrative and staff changes arising from this type of automation, but little or no scientific rethinking is required.

The economics of a conventional integrated system do not appear particularly favourable at the present time. From the aspect of systems design, data processing requires too many human stop-go interventions. If an integrated system is to be truly economic, it must be truly integrated. More important, I find automation of existing soil investigation procedures scientifically unenterprising. A very important reason for conventional techniques lies in the human inability to handle large data sets. The computer liberates us from this constraint. There is no longer an excuse for abandoning the original data on the way.

Conclusion

Current developments in integrated soil information systems have not broken away from century-old pedological tradition in anything but filing technique. There would seem to be significant arguments for reappraisal the conceptual principles behind procedures. The best choice that can be made in 1975 between conventional and alternative integrated systems, is probably to make a system that is highly modular, so that modules can be used in either approach. This will ease the use of existing data, and allow a gradual transition, also psychologically, into a more purpose-built system.

The project group to which I belong is faced with this very problem. The Netherlands is a small country facing intense pressure on the land. We are just beginning to experience the difficulties of interpretation and use of classical soil information. Parallel to the development of a conventional integrated system we are devoting considerable attention to future alternative systems. We have to-day implemented modules which allow us to proceed from the original field data to the final interpretation map without data reduction before the map drawing stage. The soil or interpretation map is constructed in the machine from point or mixed point and line data. The map is constructed from the geographical distribution of individual soil attributes, not the distribution of soil classes. When soil classification is required, numerical classification algorithms are applied. When such algorithms are not thought applicable, interactive Bayesian probability classification is attempted.

I believe that we are indeed

moving towards integrated systems for streamlining data handling. I foresee a generation of pantograph systems. But I hope an alternative approach will soon be developed, so that the mechanization of conventional systems will prove a brief interlude. I foresee that the original data will soon regain their place in soil investigation.

References

Bidwell, O.W. & F.D. Hole (1964). An experiment in the numerical classification of some Kansas soils. Proc. Soil Sci. Soc. Am. 28, 263-269

Coles, D.J. (1974). Ordnance Survey Digital Mapping. In: Automated cartography. British Cartographic Society Special Publ. 1, 69-85

De Leenheer, L., F. Appelmans & J. Vandamme (1968). Le système des cartes perforées de la section caractérisation du sol de la cartographie des sols de la Belgique. Pédologie, 18, 208-227

Dumanski, J., B. Kloosterman & S.E. Brandon (1975). Concepts, objectives and structure of the Canada Soil Information System. Can. J. Soil Sci., 55, 181-187

Flach, K.W. (1975). Report on Activities. Catalogue on Soil Information Systems, Wageningen. USA-7 (stencilled)

Jongen, P. (1975). The use of data processing methods in land development studies. Int. Symposium on the 'World Food Problem', Khartoum

Låg, J. (1962). Classification and fertility of Norwegian forest soils. Trans. Comm. IV & V int. Soc. Soil Sci., New Zealand, 540-542

Nichols, J.D. & L.J. Bartelli (1974). Computer-generated interpretive soil maps. J. Soil Water Cons., 29

ORSTOM (1969). Glossaire de Pédologie. Initiations Documentations Techniques 13. Paris

Soil Survey of England and Wales (1973). Predicted underdrainage treatment for arable land use. Abingdon. Harpenden (map)

Troedsson, T. (1964). Forest soil investigations made in Sweden in field work carried out in conjunction with the National Forest Survey. Trans. 8th int. Congr. Soil sci., Bucharest, 5, 1045-1050

Present and future role of FAO in soil information systems

J. Riquier, Soil Survey, Soil Resources Development and Conservation Service, Food and Agriculture Organization of the United Nations, Rome, Italy

Introduction

Since 1968, the FAO Land and Water Development Division (AGL) has included in its programme of work, the development of a soil information system. Since 1974, this activity has been carried out in liaison with the Environmental Data Bank project of UNEP.

A first study of a soil information system was initiated by A.W. Moore (CSIRO, Queensland, Australia), for FAO in 1969.

His work provided the basis for a complete international Soil Data Bank. Even now, little needs to be changed in this report. Unfortunately the storage and retrieval system proposed was not adapted to the computer FAO had and international collaboration was arranged before further progress could be made.

A 'Report on the Soil Data Processing System in FAO' was produced in September 1971, outlining the role of FAO in this activity.

A questionnaire has been sent to different national organizations to ascertain their interest and their degree of participation in this bank. 59 organizations have recorded their support.

A Technical Officer was appointed as FAO officer to be in charge of the Soil Information System. He undertook a travel through the United States, France, Germany, Belgium and the Netherlands to collect documentation on all existing systems, especially the Soil Data System (SDS) of US Department of Agriculture, the ORSTOM and SOGREAH systems in France, and the systems for processing of raw data by multivariate analysis, multiple regression, numerical classification in Germany and Belgium.

A meeting on 'Land Productivity Evaluation' was held in Sofia in September 1971 to consider the subject. The meeting studied methods of interpreting soil data, evaluating soil productivity, predicting yield, etc. These methods make frequently use of the computer for calculations, and these application programmes influence the choice of input data, the data files, the data retrieval system and, in general, the structure of the data bank.

An *ad hoc* Consultation on 'Computerized Soil Data Interpretation for Development Purposes' was held in Rome in April 1972. Many ideas were exchanged between 22 participants. The main recommendations resulting from the meeting were: 1) each interested institution was invited, although using its own methods of data collection, to participate in a World Data Bank; 2) FAO and other international agencies (Unesco, International Society of Soil Science) should take up the responsibility for organizing a Soil Data Processing System.

Many papers on soil data processing have been presented by FAO in different meetings from 1971 to 1974. Examples are: 'Soil

Data Processing as an Aid to Regional Land Use Planning', Land and Water Resources Evaluation, Mexico 1971, 'Approach of FAO in the Field of Soil Data Processing' in a session of the Working Party on rice soils, water and fertilizer practices, Bangkok, November 1972, and 'The Production of Soil Maps for the FAO by Automatic Methods' in an Expert Consultation on Soil Degradation held in Rome, June 1974.

FAO has also participated in the meetings: Colloque international 'Informatique et Environment', Arlon, May 1974, meetings of the 'Groupe Pédologique de l'Association Internationale Informatique et Biosphère', Nîmes and Grenoble, 1973.

During this period FAO began to develop a system for processing by computer of results from fertilizer trials.

Present role of FAO

The preceding statements show that FAO has been active in this field during the period from 1969 to the present, but before explaining achievements, let me place the project in perspective.

Almost all soil information systems in the world are at an experimental stage and progress quickly with the progress of computers. FAO cannot adopt a system without experiments. The system must also be compatible with existing systems in different countries. This explains that FAO has been playing a waiting game during recent years. It now seems the moment for FAO to coordinate all activities and to obtain international agreement, firstly on the soil data to be registered, on their codification and perhaps on the bank structure, to elaborate a common retrieval system. The 'Catalogue on Soil Information Systems', an initiative of ISSS, and the meeting of the ISSS Working Group will facilitate the comparison and the bringing together of the different systems.

The difficulties are very real: the number and the cost of personnel necessary to create a soil information service, the data collection by persons who often do not see the purpose of such collection, the difficulty of going beyond an experimental study to a routine situation, the amount of data necessary to characterize the environment or to calculate soil productivity. FAO has not yet overcome these difficulties.

Achievements

Soil Data File

The general structure of this file was established in 1971. A second revised draft of items definitions has now been prepared.

Data entry has been the object of much thought and discussion, giving particular consideration to the constraints of data capture (which should be as simple as possible) and the constraints of data interchange between data banks (which should allow each data bank to handle the data according to its own storage and processing system). For data interchange, the proposed solution is a standard card image on tape, each entry bearing an initial identification number. This common communication format for the exchange of data between international users, has been discussed with Interorganization Board of International Computer Centre of UN in Geneva, and the FAO Computer Systems Branch in FAO.

Compatibility between different soil data systems have been studied simultaneously with the establishment of the soil data file.

A 'Guideline to Codifying Soil Data' has been prepared but not yet published. It will appear before the end of 1975 in provisional form and will be widely distributed for comment. Our aim is to obtain the consensus of as many experts as possible on this system, which is intended to become international.

The publication will list the items and variables to describe a soil profile, together with physical and chemical data and environmental characteristics. Many other soil information systems already adopted by other institutions, organizations or countries have been compiled in the listing and we have chosen the common soil properties currently collected throughout the world with the exception of properties particular to an ecological zone, or for special use. Such cases can be handled in the regional, national or local banks; an example is the permafrost horizon description. The objective has been to codify and standardize the entry format of items common to the majority of banks. The selection of items has been governed by using those required for soil classification or for land evaluation and soil productivity.

Cross references with the climatic bank of WMO, with the FAO land use and geographic data file and with the FAO farm management, are foreseen in the input cards.

Hayatsu of the FAO Computer Branch has recently proposed an appropriate storage, retrieval and updating system, which is adapted to the FAO computer, IBM 370.

Geographic Data File and Automated Cartography

The data on soil, environment and land use, are grouped and linked to a mapping unit of the FAO/ Unesco Soil Map of the World. Later they will be linked to an agro-ecological map.

This programme has been carried out under contract with the Experimental Cartography Unit (ECU) at the Royal College of Art in London. It is as yet experimental, but shows promise. It will probably be used to automatically draft a land degradation map, from the World Soil Map.

The South America sheet of the Soil Map of the World has been digitized, and data on climate, physiography, geology, vegetation, land use, land management (two to eight punched cards for each mapping unit) are recorded by soil units, i.e. the soil association comprising the mapping unit. Retrieval of 1-20 single or associated land characteristics can be achieved by the computer and the areas with these characteristics can be displayed on a cathode ray tube and photographed, or drafted on a plotting table on completion of editing. Changes of map scale and changes of projection are possible.

Computerized contour maps can also be drafted from data points by a computer programme of ECU. FAO has used this programme to produce a contour map of equal climatic aggressivity related to erosion, from climatic stations for each of which an aggressivity index was calculated. This experimental work will be very useful, in future, to formulate a land degradation map.

The cartographic file and automated cartography are very important for all FAO work on environment.

Data recording, in these activities, is unfortunately laborious both for the land characteristics and digitization of the area boundaries (numerous errors of digitizing must be corrected). Additionally these programmes are costly, particularly in the experimental stage.

Fertilization File and Bank
The FAO Fertilizer Programme puts out, each year, a large amount of data on field fertilizer demonstration and trials results. Fertilizer response curves and the optimum economic rate of application can be obtained from this data. A sophisticated set of computer programmes for these calculations (by Colwell) have recently been published by FAO. A programme, using the orthogonal polynomial method, permits the calculation of economic rate, not only with the interaction of other fertilizers but also of other productivity factors such as climate, soil, percentage nutrients, soil texture, etc. This programme requires special trial arrangements.

FAO now routinely uses a simplified programme determining the optimum economic rate of fertilizers. Interactions between fertilizers are taken in account in the trials but not in the mathematical equation. A great advantage of this particular computer programme is that it permits stratification of results according to climate, soil type or some physical or chemical soil properties, divided in classes. Amounts of fertilizers can be recommended as a function of these different criteria. The results are recorded on magnetic tape and can be retrieved by country, soil, climate, and updated. This is the basis of the FAO Fertilization Bank.

Land-Use File
This activity is still in the project stage. Ultimately it will obviously have to be perfectly correlated with the Soil Data File and with the farm management data collection, analysis, storage and retrieval system of the FAO Agriculture Division. This farm management system records yield, crop, soil, climate, limiting factors, land tenure, agricultural practices and management by parcel of land or by farm.

Let me stress the fact that it is a land-use file and not a soil-use file. The use of soil is already recorded in the Soil Data File. The Land-Use File is the nearest to a geographic data file, because it concerns areas.

Miscellaneous Computer Programmes
For some particular purposes, FAO uses general packages of IBM or programmes especially written for the particular purpose. Attempted correlation of fertilizer response with soil analysis, by a multiple regression programme written by an FAO programmer, is an example.

The calculation of soil productivity from plant, soil and climate data is another example. This is in an experimental stage and employs a computer programme PARME (parametric method) written by Culot and me. This activity is a good example of application programmes, which use soil data, and climatic data, and will use plant physiological data, extracted from the corresponding banks.

Linear programming has also been used for an FAO project in Korea to optimize peasant incomes, by crops best suited to various soil types.

Future role of FAO

FAO is less interested in theoretical research programmes on a particular point than in large application programmes having a practical aim.

Many computer programmes have

not a sufficiently large application to cover world problems. Such programmes require special data which are not 'normally' available and must be specially collected.

However FAO has to collect these programmes for use in its field projects. The ISSS catalogue is a good basis which must be widened to particular agricultural problems: plant physiology, hydrology, plant-climate relationship, water balance in soils, erosion in watersheds, etc. For its own staff FAO needs programmes of worldwide use, taking their data from worlwide banks or regional banks having the same soil information system.

The present FAO objectives on this matter are as follows:
- To promote local or regional banks on soil, land use, geographical data;
- To publish a list of items, a standard codification, a storage and retrieval system able to adapt itself to different situations and constituting a common basis to local banks. A choice of files must be previously effected. Exchange between banks will be reduced to magnetic tape exchanges.
- To coordinate the soil information system with other banks, such as climatic, and plant physiology, by cross reference, in collaboration with UNEP.
- To create an international soil bank gathering the data of local or regional banks. It appears that the international bank should not only be a bank of references to local or regional banks, but should additionally record all the data.
- To develop the application programmes, which must justify the bank's existence.

The following points and difficulties are pertinent to this programme of activities:
1. It is difficult to gain acceptance of a soil information system in countries that have already developed their own system, perhaps better adapted to local conditions and needs. In consequence we think that it is necessary to act quickly to avoid the multiplication of different systems.
2. The choice of a soil information system implies the use of a field record form to be completed by experts in the field. It is difficult to write a practical model, suitable for different conditions.
3. Cross-references between different files save time and money, provided the other files can be interrogated for each question. It is proposed that each cross-reference can provide total information but that usually each file will contain sufficient data for the immediate objective.
4. Until now computer programmes in FAO, mainly in experimental stage, have been costly and time-consuming. We hope that it will not be the same in the future. A budget for the creation of a specialized service has yet to be made.
5. A soil data bank becomes useful when sufficient data are recorded. To collect data, we are dependent on the good will of soil specialists. To process the data, for instance statistically, we must have a sufficiently large collection of data on a given attribute.

It is necessary however to establish a code for the future and FAO will submit, very soon, an international bank project for the comments of experts. The discussion at this meeting will very much help in finalizing the text for this project.

Participants to the meeting of the working group on soil information systems, 1 - 4 Sept. 1975

Belgium

Mr F. Delecour
Centre d'Etude des Sols
Forestiers de la Haute Belgique
(I.R.S.I.A.)
Faculté Agronomique
B-5800 GEMBLOUX

Mr G. Rousseaux
Fac. Sci. Agronomiques de l'Etat
59 Avenue de la Faculté
B-5800 GEMBLOUX

Prof. A. van Wambeke
Rijksuniversiteit Gent
Rozier 44
B-9000 GENT

Bulgaria

Prof. I.P. Garbouchev
N.Poushkarov Inst. of Soil Science
Shosse Bankya 5
SOFIA-24

Dr A.N. Sadovski
N.Poushkarov Inst. of Soil Science
Shosse Bankya 5
SOFIA-24

Canada

Dr T.M. Gordon
Geological Survey of Canada
601 Booth Street
OTTAWA, Ontario K1A OE8

Dr B. Kloosterman
Soil Research Inst. Canada Dept Agric.
Central Experimental Farm
OTTAWA, Ontario K1A OC6

Ms G.M. Martin
Geological Survey of Canada
601 Booth Street
OTTAWA, Ontario K1A OE8

Dr R.F. Tomlinson
IGU Commission on Geographic Data Sensing and Processing
226 O'Connor Street
OTTAWA, Ontario K2P IT9

France

Mr G. Auxenfants
17, Avenue des Écoles
F-91320 WISSOUS

Mr N.C. Bang
GERDAT B.P.5035
F-34032 MONTPELLIER CEDEX

Mr D. Chabanel
Science du Sol, NESA
65 Rue de St.Brieuc
F-35042 RENNES CEDEX

Ms M. Dosso
Science du Sol, ENSA
65 Rue de St.Brieuc
F-35042 RENNES CEDEX

Ir R. van den Driessche
ORSTOM Soil Data Base
70 route d'Aulnay
F-93140 BONDY

Dr A. García Gómez de Van den Driess
ORSTOM Remote Sensing
70 route d'Aulnay
F-93140 BONDY

France (continued)

Dr M.C. Girard
Institut National Agronomique
Paris-Grignon
F-78850 THIVERVAL-GRIGNON

Mr P. Jongen
c/o SOGREAH
47, Av. Marie-Reynoard
F-38100 GRENOBLE

Mr J. Kilian
Inst. de Recherches Agronomiques
Tropicales et des Cultures Vi-
vrieères
110, rue de l'Université
F-75340 PARIS CD 07

Dr K. Lange
UNESCO
7 Pl. de Fontenoy
F-75700 PARIS

Mr J.P. Legros
INRA-Science du Sol
ENS Agronomique
F-34060 MONTPELLIER

Dr R. Maignien
ORSTOM Service Scientifiques
Centraux
24 Rue Bayard
F-75008 PARIS

Prof. A. Ruellan
Science du Sol, ENSA
65 Rue de St. Brieuc
F-35042 RENNES CEDEX

Germany

Dr J. Lamp
Inst. für Pflanzenernährung und
Bodenkunde
Olshausenstr. 40-60
D-23 KIEL

Dr K.H. Oelkers
Niedersächsisches Landesamt für
Bodenforschung
Postfach 230153
D-3 HANNOVER 23

Ghana

Mr F. Ablorh
c/o Dept. of Agricultural
Science
Oxford University, Parks Rd.,
OXFORD, OX1 3PF (U.K.)

Israel

Mr Y. Peleg
Ministry of Agriculture
Div. of Mapping and Soil Survey
Yehuda Hayamit St.35
YAFFO P.O.Box 15009

Italy

Mr H. Hayatsu
FAO
Via delle Terme di Caracalla
I-00100 ROMA

Mr J. Riquier
FAO
Via delle Terme di Caracalla
I-00100 ROMA

Mr N.H. Stålbrand
FAO
Via delle Terme di Caracalla
I-00100 ROMA

Netherlands

Mr J.T.J. Beeren
Adviesbureau Arnhem
Oeverstraat 120
ARNHEM

Dr S.W. Bie
Soil Survey Institute
P.O.Box 98
WAGENINGEN

Prof.Dr Ir P. Buringh
Agricultural University
Dept. of Tropical Soil Science
Duivendaal 10
WAGENINGEN

Netherlands (continued)

Dr Ir Th.J. Ferrari
Institute of Soil Fertility
Verl. Oosterweg 92
HAREN (Gr)

Ir J.J. de Gruijter
Soil Survey Institute
P.O.Box 98
WAGENINGEN

Dr Ir J. Schelling
Soil Survey Institute
P.O.Box 98
WAGENINGEN

Ir L. van Zuijlen
Int. Institute for Aerial Survey
and Earth Sciences (ITC)
P.O.Box 6
ENSCHEDE

New Zealand

Mr D.J. Giltrap
c/o Dept. of Agricultural Science
Oxford University, Parks Rd.
OXFORD OX1 3PF (U.K.)

Rumania

Dr D. Teaci
Research Institute for Soil
Science and Agrochemistry
Bd Marasti 61
7000-BUCHAREST

Spain

Dr A. Guerra Delgado
de Edafología y Biología Vegetal
Serrano, 115 dplo
MADRID-6

Dr F. Monturiol Rodriguez
de Edafología y Biología Vegetal
Serrano, 115 dplo
MADRID-6

Sweden

Prof.Dr T. Troedsson
Dept. of Forest Soils
Royal College of Forestry
S-75007 UPPSALA

Switzerland

Dr K. Peyer
Swiss Federal Research Station
for Agronomy Zürich-Reckenholz
P.O.Box CH-8046
ZÜRICH

United Kingdom

Dr D.F. Ball
Institute of Terrestrial Ecology
Bangor Research Station
Penrhos Road
BANGOR, Gwynedd

Dr P.H.T. Beckett
Dept. of Agricultural Science
Oxford University
Parks Road
OXFORD OX1 3PF

Ms J.E.R. Hoare
Dept. of Agricultural Science
Oxford University, Parks Road
OXFORD OX1 3PF

Mr R.H.E. Inkson
Macaulay Institute for Soil
Research
Craigiebuckler
ABERDEEN AB9 2QJ

Dr K.G. Jeffery
Institute of Geological Sciences
(Computer Unit)
Exhibition Road
LONDON SW7 2DE

Mr D.M. Lang
Land Resources Division
Tolworth Tower
SURBITON, Surrey

Mr J.M. Ragg
Soil Survey of Scotland
The Edinburgh School of Agri-
culture
West Mains Road
EDINBURGH EH9 3JG

Dr R. Webster
Soil Survey of England and Wales
ARC, Weed Research Organization
Sandy Lane
Yarnton, OXFORD OX5 1PF

USA

Dr G.L. Decker
USDA, Soil Conservation Service
BOZEMAN, Montana

Dr K.W. Flach
USDA, Soil Conservation Service
Rm. 5207 South Building
WASHINGTON D.C. 20250

Mr C.G. Johnson
USDA, Soil Conservation Service
6505 Belcrest Rd.
HYATTSVILLE, Maryland 20782

Yugoslavia

Dr M. Ćirić
Sumarski Fakultet
Zagrebacka 20
71000-SARAJEVO

Venezuela

Mr J.A. Comerma
Seccion Suelos; Ceniap; M.A.C.
MARACAY